THE SELF-AWARE UNIVERSE

ALSO BY AMIT GOSWAMI

The Concepts of Physics
Quantum Mechanics

With Maggie Goswami

The Cosmic Dancers

THE SELF-AWARE UNIVERSE

HOW CONSCIOUSNESS CREATES THE MATERIAL WORLD

Amit Goswami, Ph.D.

WITH RICHARD E. REED
AND MAGGIE GOSWAMI

A Jeremy P. Tarcher / Putnam Book
PUBLISHED BY G. P. PUTNAM'S SONS
New York

To my brother the philosopher
Nripendra Chandra Goswami

Credits appear on page 320.

A Jeremy P. Tarcher/Putnam Book
Published by G. P. Putnam's Sons
Publishers Since 1838
200 Madison Avenue
New York, NY 10016

Published simultaneously in Canada

Library of Congress Cataloging-in-Publication Data

Goswami, Amit.
The self-aware universe : how consciousness creates the material world / Amit Goswami with Richard E. Reed and Maggie Goswami.—

p. cm.
Includes bibliographical references and index.
ISBN 0-87477-669-4 (alk. paper)
1. Quantum theory. 2. Physics—Philosophy. 3. Science—Philosophy. 4. Religion and science. I. Reed, Richard E. II. Goswami, Maggie. III. Title.
QC174.13.G67 1993 92-32178 CIP
530.1'2—dc20

Design by Irving Perkins Associates, Inc.

Printed in the United States of America
3 4 5 6 7 8 9 10

This book is printed on acid-free paper.
∞

CONTENTS

Part 3

Self-Reference: How the One Becomes Many

Part 4

The Re-enchantment of the Person

ILLUSTRATIONS

PREFACE

When I was a graduate student studying quantum mechanics, a group of us would spend hours discussing such esoterica as, Can an electron really be at two places at the same time? I could accept that, yes, the electron can be at two places at the same time; the message of quantum mathematics, although full of subtlety, is unambiguous on this point. Does an ordinary object, however—a chair or a desk, things that we call "real"—behave like an electron? Does it become a wave and start spreading in the wave's inexorable way whenever no one is looking?

Objects found in our everyday experience do not seem to behave in the strange ways common to quantum mechanics. Thus, subconsciously, it is easy for us to be lulled into thinking that macroscopic matter is different from microscopic particles—that its conventional behavior is governed by Newtonian laws, which are referred to as classical physics. Indeed, many physicists stop puzzling over the paradoxes of quantum physics and succumb to this solution. They divide the world into quantum and classical objects—and so did I, although I did not realize what I was doing.

To forge a successful career in physics, you cannot worry too much about such recalcitrant questions as the quantum puzzles. The pragmatic way of doing quantum physics, I was told, is to learn to calculate. I therefore compromised, and the tantalizing questions of my youth gradually shifted to a back burner.

They did not, however, disappear. Circumstances shifted for me, and—after my umpteenth bout of the stress heartburn that characterized my competitive-physics career—I began to remember the exuberance I once felt about physics. I realized that there must be a joyful way of approaching the subject, but I needed to restore my spirit of inquiry into the meaning of the universe and to abandon the mental compromises I had made for career motives. A book by the philosopher Thomas Kuhn that distinguishes paradigm research from scientific revolutions that shift paradigms was very helpful. I had done my share of paradigm research; it was time to move on to the frontier of physics and to think about a paradigm shift.

Just about the time of my personal crossroads, Fritjof Capra's book *The Tao of Physics* came out. Although my initial reaction to the book was jealousy and rejection, it did touch me deeply. After a while I could see that the book broaches a problem that it does not investigate thoroughly. Capra delves into the parallels between a mystical view of the world and that of quantum physics but does not investigate the reason for these parallels: Are they more than coincidence? At last, I had found the focus of my inquiry into the nature of reality.

Capra's entree to questions about reality was through elementary particle physics, but I intuited that the key issues are most directly confronted in the problem of how to interpret quantum physics. This is what I set out to investigate. I did not anticipate initially that this would be such an interdisciplinary project.

I was teaching a course on the physics of science fiction (I have always had a soft spot for science fiction), and a student commented: "You talk like my psychology professor, Carolin Keutzer!" A collaboration with Keutzer ensued that, although not leading to any major insight, did introduce me to a lot of relevant psychological literature. I eventually became familiar with the work of Mike Posner and his cognitive psychology group at the University of Oregon, which was to play a crucial role in my research.

Besides psychology, my subject of research demanded considerable knowledge of neurophysiology—brain science. I met my neurophysiology teacher through the mediation of John Lilly, the famous dolphinologist. Lilly had kindly invited me to participate in a week-long Esalen seminar that he was giving; Frank Barr, M.D., was also a participant. If my passion was quantum mechanics, Frank's was brain theory. I was able to learn from him just about everything I needed to begin the brain-mind aspect of this book.

One other crucial ingredient for my ideas to gel consisted of the theories of artificial intelligence. Here, too, I was very fortunate. One of the exponents of artificial intelligence theory, Doug Hofstadter, began his career as a physicist; he earned his degree at the University of Oregon graduate school, where I teach. Naturally, when his book came out, I had a special interest in it and learned some of my key ideas from Doug's research.

The meaningful coincidences go on and on. I was initiated to the research in parapsychology through many discussions with another of my colleagues, Ray Hyman, who is a very open-minded skeptic. Last but not least of the important coincidences was my meeting with three mystics in Lone Pine, California, during the summer of 1984: Franklin Merrell-Wolff, Richard Moss, and Joel Morwood.

In a sense, since my father was a Brahmin guru in India, I grew up immersed in mysticism. At school, however, I started a long detour through the conventional training and practice of a scientist with a compartmentalized specialty. This direction pointed me away from my childhood sympathies and resulted in my believing that the objective reality defined by conventional physics is the only reality—anything subjective is due to a complex dance of atoms waiting to be deciphered by us.

In contrast, the Lone Pine mystics talked about consciousness as being "original, self-contained, and constitutive of all things." Their ideas led to considerable cognitive dissonance for me in the beginning, but eventually I realized that one can still do science even if one assumes the primacy of consciousness rather than of matter. This way of doing science, moreover, routs not only the quantum paradoxes of my teenaged puzzling but also new ones of psychology, the brain, and artificial intelligence.

Well, this book is the end product of my roundabout journey. It took ten to fifteen years to overcome my bias for classical physics and then to research and write the book. I hope that the fruit of my effort will be worth your while. To paraphrase Rabindranath Tagore,

I have listened
And I have looked
With open eyes.
I have poured my soul
Into the world
Seeking the unknown
Within the known.
And I sing out loud
In amazement.

Obviously, many more people than the aforementioned contributed to the book: Jean Burns, Paul Ray, David Clark, John David

Garcia, Suprokash Mukherjee, the late Fred Attneave, Jacobo Grinberg, Ram Dass, Ian Stuart, Henry Stapp, Kim McCarthy, Robert Tompkins, Eddie Oshins, Shawn Boles, Fred Wolf, and Mark Mitchell—just to mention a few. The encouragement and emotional support of friends were important, notably from Susanne Parker Barnett, Kate Wilhelm, Damon Knight, Andrea Pucci, Dean Kisling, Fleetwood Bernstein, Sherry Anderson, Manoj and Dipti Pal, Geraldine Moreno-Black and Ed Black, my late colleague Mike Moravcsik, and especially our late and beloved friend Frederica Leigh.

Special thanks go to Richard Reed, who convinced me to submit the manuscript for publishing and who took it to Jeremy Tarcher. In addition, Richard has given important support, critique, and help with the editing. Of course, my wife, Maggie, has contributed so much both to the development of ideas and to the language that expresses the ideas that this book literally would have been impossible without her. The editors provided by J. P. Tarcher, Inc.—Aidan Kelly, Daniel Malvin, and especially Bob Shepherd—have earned my heartfelt thanks, as has Jeremy Tarcher himself for believing in this project. I thank you all.

FOREWORD

It wasn't that long ago when we physicists believed that we had finally come to the end of all our searching: We had reached the end of the road and found the mechanical universe perfect in all of its splendor. Things behave the way they behave because they were the way they were in the past. They will be the way they will be because they are the way they are, and so on. Everything fit in a nice tiny package of Newtonian-Maxwellian thought. There were mathematical equations that actually fit the behavior of nature. There was a one-to-one correspondence between a symbol on the page of the scientific paper and the movement of the tiniest to the grossest object in space and through time.

It was the end of a century, the nineteenth, to be exact, and the renowned A. A. Michelson, speaking about the future of physics, said that it would consist of "adding a few decimal places to results already obtained." To be fair, Michelson believed he was quoting the famous Lord Kelvin in making this remark. Actually it was Kelvin who said that indeed everything was perfect in the landscape of physics except for two dark clouds obscuring the horizon.

These two dark clouds, it turned out, not only blotted out the sun of the Turneresque, Newtonian landscape, they changed it into a bewildering abstract Jackson Pollock vision of points, smears, and waves. These clouds were the forerunners of the now-famous quantum theory of everything.

Thus here we are again at the end of the century, the twentieth, to be exact, and once more clouds are gathering to obscure the landscape of even the quantum world of physics. Just as before, the Newtonian landscape certainly had and still has its admirers. It still works in explaining a vast range of mechanical phenomena, from spaceships to automobiles, from satellites to can openers; and yet, just as the quantum abstract painting ultimately has shown that this Newtonian landscape is made up of seemingly random dots (quanta), there are still many of us who believe that ultimately there is some kind of objective mechanical order underlying everything, even the quantum dots.

Science, you see, proceeds by a very fundamental assumption of the way things are or must be. That assumption is the very thing that Amit Goswami, with the assistance of Richard E. Reed and Maggie Goswami, brings into question in the book you are about to read. For this assumption, like its cloudy predecessors of the century before, seems to be signaling not only the end of a century but the end of science as we know it. That assumption is that there exists, "out there," a real, objective reality.

This objective reality is something solid; it is made up of things that have attributes, such as mass, electrical charge, momentum, angular momentum, spin, position in space, and continuous existence through time expressed as inertia, energy, and going even deeper into the microworld, such attributes as strangeness, charm, and color. And yet the clouds still gather. For in spite of all that we know about the objective world, even with its twists and turns of space into time into matter, and the black clouds called black holes, with all of our rational minds working at full steam ahead, we are still left with a flock of mysteries, paradoxes, and puzzle pieces that simply do not fit.

But we physicists are a stubborn lot, and we fear the proverbial toss of the baby out with the bathwater. We still lather and shave our faces watching carefully as we use Occam's razor to make sure that we cut away all superfluous "hairy assumptions." What are these clouds that obscure the end of the twentieth century's abstract art form? They boil down to one sentence: The universe does not seem to exist without a perceiver of that universe.

Well, at some level this certainly makes sense. Even the word "universe" is a human construct. So it would make some kind of sense that what we call the universe depends on our word-making capacity as human beings. But is this observation any deeper than a simple question of semantics? For example, before there were human beings, was there a universe? It would seem that there was. Before we discovered the atomic nature of matter, were there atoms around? Again, logic dictates that the laws of nature, forces and causes, etc., even though we didn't know about such things as atoms and subatomic particles, certainly had to exist.

But it is just these assumptions about objective reality that have been called into question by our present understanding of physics. Take, for example, a simple particle, the electron. Is it a little speck of matter? It turns out that to assume that it is such, consistently behaving itself as such, is clearly wrong. For at times it appears to be a cloud made up of an infinite number of possible electrons that "appear" as a single particle when and only when we observe one. Furthermore, when it is not a single particle it appears to be an undulating wavelike cloud that is capable of moving at speeds in excess of light speed, totally contradicting the Einstein concern that nothing material can move faster than light. But Einstein's worry is assuaged, for when it moves this way, it is not actually a piece of matter.

Take as another example the interaction between two electrons. According to quantum physics, even though the two electrons may be vast distances apart, the results of observations carried out upon them indicate that there must be some connection between them that allows communication to move faster than light. Yet before those observations, before a conscious observer made up his or her mind, even the form of the connection was totally indeterminate. And as a third example, a quantum system such as an electron in a bound physical state appears to be in an indeterminate state, and yet the indeterminacy can be analyzed into component certainties that somehow add to the original uncertainty. Then along comes an observer who, like some gigantic Alexander chopping the Gordian knot, resolves the uncertainty into a single, definite but unpredictable state simply by observing the electron.

Not only that, the blow of the sword could come in the future determining what state the electron is in now. For we have now even the possibility that observations in the present legitimately determine what we can say was the past.

Thus we have come to the end of a road once again. There is too much quantum weirdness around, too many experiments showing that the objective world—one that is running forward in time like a clock, one that says action at a distance, particularly instantaneous action at a distance, is not possible, one that says a thing cannot be in two or more places at the same time—is an illusion of our thinking.

So what can we do? This book may have the answer. The author posits a hypothesis that is so strange to our Western minds as to be automatically dismissed as the ravings of an Eastern mystic. It says that all of the above paradoxes are explainable, are understandable, if we are to give up that precious assumption that there is an objective reality "out there" independent of consciousness. It says even more, that the universe is "self-aware" and that it is consciousness itself that creates the physical world.

As Goswami uses the word "consciousness," he is implying something perhaps more profound than you or I would imply. In his terms consciousness is something transcendental—outside of space-time, nonlocal, and all-pervading. It is the only reality, yet we are able to glimpse it only through the action that gives rise to the material and mental aspects of our observational processes.

Now, why is this so hard for us to accept? Perhaps I am presuming too much to say that it is hard to accept for you the reader. Perhaps you find this hypothesis self-evident. Well, at times I am comfortable with this, but then I bump into a chair and bruise my leg. That old reality sinks in, and I "see" myself distinct from the chair as I curse its position in space so arrogantly separate from mine. Goswami addresses this issue admirably and provides several often amusing examples to illustrate his thesis that I and the chair arise out of consciousness.

Goswami's book is an attempt to bridge the age-old gap between science and spirituality, which he believes his hypothesis accomplishes. He has much to say about monistic idealism and how it alone resolves the paradoxes of quantum physics. Next he looks into the age-old question of mind and body or mind and brain and shows how his overarching hypothesis that consciousness is everything heals the Cartesian split—and in particular, in case you were wondering, even how one consciousness appears to be so many separate consciousnesses. Finally, in the last part of the book he offers a glimmer of hope as we grope through the clouds to the twenty-first century as he explains how this hypothesis will actually accomplish the re-enchantment of the person with his environment, something we assuredly need. He explains how he experienced his own theory when he realized the mystical truth, the "nothing-but-consciousness must be experienced in order to be truly *understood*."

Reading this book, I also began to feel this. Given that the hypothesis is truth, then it would follow that you too will have this experience.

Fred Alan Wolf, Ph.D.
author of *The Dreaming Universe*,
Taking the Quantum Leap, and other books
La Conner, Washington

PART 1

THE INTEGRATION OF SCIENCE AND SPIRITUALITY

A critical level of confusion permeates the world today. Our faith in the spiritual components of life—in the vital reality of consciousness, of values, and of God—is eroding under the relentless attack of scientific materialism. On the one hand, we welcome the benefits derived from a science that assumes the materialist worldview. On the other hand, this prevailing worldview fails to satisfy our intuitions about the meaningfulness of life.

During the past four hundred years, we have gradually adopted the belief that science can be built only on the notion that everything is made of matter—of so-called atoms in the void. We have come to accept materialism dogmatically, despite its failure to account for the most familiar experiences of our daily lives. In short, we have an inconsistent worldview. Our predicament has fueled the demand for a new paradigm—a unifying worldview that will integrate mind and spirit into science. No new paradigm, however, has surfaced.

This book proposes such a paradigm and shows how we can develop a science that embraces the religions of the world, working in concert with them to understand the whole human condition. The centerpiece of this new paradigm is the recognition that modern science validates an ancient idea—the idea that consciousness, not matter, is the ground of all being.

The first part of this book introduces the new physics and a modern version of the philosophy of monistic idealism. On these two pillars, I shall attempt to construct the promised new paradigm, a bridge over the chasm between science and religion. Let there be commerce between the two.

Chapter 1

THE CHASM AND THE BRIDGE

I SEE A STRANGE, torn-up caricature of a man beckoning to me. What is he doing here? How can he exist in so fragmented a state? What do I call him?

As if reading my mind, the tortured figure speaks: "In my condition, what difference does a name make? Call me Guernica. I am looking for my consciousness. Am I not entitled to consciousness?"

I recognize the name. *Guernica* is the masterpiece Pablo Picasso painted in protest against the Fascist bombing of a little Spanish town of that name.

"Well," I reply, trying to soothe him, "if you will tell me precisely what you need, perhaps I can help."

"You think so?" His eyes light up. "Maybe you will plead my case?" He looks at me yearningly.

"With whom? Where?" I ask, intrigued.

"Inside. They are having a party while I am abandoned out here unconscious. Maybe if I find my consciousness, I'll be whole again."

"Who are they?" I ask.

"The scientists, the ones who decide what's real."

"Oh? The situation can't be so bad then. I am a scientist. Scientists are an open bunch. I'll go talk to them."

The people at the party are divided into three separate groups like the islands of the Bermuda triangle. I hesitate for a moment, then stride firmly toward one of these groups—when in Rome and all that. The conversation is intense. They are talking about quantum physics. They must be physicists.

"Quantum physics gives predictions for the events that we observe experimentally, nothing more," a distinguished-looking gentleman with just a touch of gray in his hair says. "Why make unsupported assumptions about reality when talking about quantum objects?"

"Aren't you a little tired of that line? A whole generation of physicists seems to have been brainwashed into thinking that an adequate philosophy of quantum physics was developed sixty years ago.[1] That is just not the case. Nobody understands quantum mechanics," says another, whose sad demeanor is obvious.

Those words scarcely register in the discussion when another gentleman, displaying an unruly beard, says with arrogant authority: "Look, let's set the context right. Quantum physics says that objects are represented by waves. Objects are waves. And waves, as we all know, can be at two (or more) places at the same time. But when we observe a quantum object, we find it all at one place, here, not over there, and certainly not both here and over there at the same time."

The bearded fellow is waving his hands excitedly. "So what does this mean in simple terms? You," he says, looking at me, "what do you say, sir?"

I am taken aback for a moment by this challenge but speedily recover. "Well, it seems that our observations, and thus we, have a profound effect on quantum objects."

"No. No. No," my interrogator thunders. "When we observe, there is no paradox. When we don't, the paradox of the object being in two places at the same time returns. Obviously, the way to avoid the paradox is to vow never to talk about an object's whereabouts in between observations."

"But what if we, our consciousness, really have a profound effect on quantum objects?" I persist. Somehow, it seems to me, Guernica's consciousness has something to do with this speculation.

"But that means mind over matter," all the people in the group cry in unison, looking at me as if I have uttered heresy.

"But, but," I stammer, refusing to be daunted, "suppose there is a way of reconciling mind over matter."

I tell them about Guernica's predicament. "Look, fellows, you have a social responsibility here. You have known for sixty years that the conventional, objective way of doing physics does not work with quantum objects. We get paradoxes. Yet you pretend objectivity, and the rest of the society misses a chance of recognizing that we—our consciousness—are intimately connected with reality. Can you imagine the impact on the ordinary person's worldview if physicists plainly admit that we are not separate but, instead, are the world and must take responsibility for it? Maybe then only, Guernica, nay, all of us can return to wholeness."

The distinguished gentleman intervenes. "I will admit when it's deep in the night and nobody's around, I have doubts. Maybe we are missing a chance. But my mother taught me, when in doubt, it's much better to pretend ignorance. We don't know a thing about consciousness. Consciousness belongs to psychology, to those guys over there," he gestures toward a corner.

"But," I persist doggedly, "suppose we define consciousness as the agency that affects quantum objects to make their behavior sensible. I am sure psychologists would consider that possibility if you guys join me. Let's take a crack at changing our separatist worldview right now." I have become certain that Guernica's chance to gain consciousness depends on the success of my rallying these people.

"It is opening Pandora's box to say that consciousness causally affects atoms. That would turn objective physics upside down; physics would not be self-contained, and we would lose our credibility." There is a tone of finality in the voice that speaks. Somebody else with a voice that I had heard before, says: "Nobody understands quantum mechanics."

"But I promised Guernica that I would plead for his consciousness! Please, hear me out," I protest, but nobody pays any attention. I have become a nonentity in this group—a nonconsciousness, like Guernica.

I decide to try the psychologists. I recognize them by the cluster of rat cages and computers in their corner.

A competent-looking woman is explaining something to a young man. "By assuming that the brain-mind is a computer, we hope to go beyond the behaviorist rat race. The brain is the computer's hardware. There is nothing, really, but the brain; that's what's real. However, the states of the brain's hardware, over time, carry out independent functions, like computer software. It's these states of the hardware that we call the mind."

"Then what is consciousness?" probes the young man.

Hey, what perfect timing. That's just what I came over here to find out—how psychologists think of consciousness! They must be the ones who have control over Guernica's consciousness.

"Consciousness is like the central processing unit, the command center of the computer," answers the woman patiently.

Her questioner, not satisfied with this reply, presses on: "If we can explain, even in principle, all our input-output performances in terms of the activity of computer circuits, then consciousness seems to be absolutely unncessary."[2]

I cannot restrain myself. "Please don't give up on consciousness yet. My friend Guernica needs it." I tell them about Guernica's problem.

Sounding like an echo of my erstwhile physicist friend, a nattily dressed gentleman casually interjects: "But cognitive psychology is not ready for consciousness yet.[3] We don't even know how to define it."

"I can give you a physicist's definition of consciousness. It has to do with the quantum."

That last word gets their attention. First, I explain about quantum objects being waves that spread in existence at more than one place and how consciousness may be the agency that focuses the waves so we can observe them at one place. "And this is the solution to your problem," I offer. "You can take the definition of consciousness from physics! And then maybe you can help Guernica."

"But aren't you mixing things up? Don't physicists say that everything is made of atoms—quantum objects. If consciousness is also made of quantum objects, how can it causally act on them? Think, man, think."

I am panicking a little bit. If these psychologists know what they are talking about, even my consciousness is an illusion, let alone Guernica's. But the psychologists are right only if all things, including consciousness, really are made of atoms. Suddenly, another possibility flashes to my mind! And I blurt out: "You are doing it all wrong! You can't be sure if all things are made of atoms—it's an assumption. Suppose all things, including atoms, are made of consciousness, instead!"

My listeners seem stunned. "Look, there are some psychologists who think that way. I admit, yours is an interesting possibility. But it is not scientific. If we want to elevate psychology to the status of science, we must keep away from consciousness—especially the notion that consciousness might be the primary reality. Sorry, fella." The woman who has spoken actually sounds quite sympathetic.

But I still haven't made any headway for Guernica's consciousness. In desperation, I turn to the last group—the third apex of the triangle. They turn out to be neurophysiologists (brain scientists). Perhaps they are the judges who really count.

The brain scientists are also having an argument about consciousness, and my expectations rise. "Consciousness is a causal entity that brings meaning to existence, I give you that," says one of them, addressing an older man who is quite thin. "But it must be an emergent phenomenon of the brain, not separate from it. After all, everything is made of matter; that's all there is."[4]

The thin fellow, speaking with a British accent, objects. "How can something made of something else act causally on what it's made of? That would be like a television commercial repeating itself by acting on the electronic circuitry of the television set. God forbid! No, consciousness has to be a separate entity from the brain in order to have a causal effect on it. It belongs to a separate world outside the material world."[5]

"But then how do the two worlds interact? A ghost cannot act on a machine."

Rudely interrupting, a third man, wearing his hair in a ponytail, laughs and says: "Both of you are talking humbug. All your problems arise from trying to find meaning in an inherently meaningless material world. Look, the physicists are right when they say there is no meaning, there is no free will, and everything is the random play of atoms."

The British supporter of a separate world for consciousness, now sarcastic, retorts: "And you think what you say makes sense! You, yourself, are the play of the random, meaningless motion of atoms, yet you make theories and think that your theories mean something."

I wedge myself into the debate. "I know a way to have meaning even in the play of atoms. Suppose that instead of everything being made of atoms, everything is made of consciousness. What then?"

"Where did you get that idea?" they challenge.

"Quantum physics," I tell them.

"But there is no quantum physics at the macro level of the brain," they all exclaim with authority, unified in their objection. "The quantum is for the micro, for the atoms. Atoms make up molecules, molecules make up cells, and cells make up the brain. We work with the brain every day; there is no need to invoke the quantum mechanics of atoms to explain the gross-level behavior of the brain."

"But you don't claim complete understanding of the brain, do you? The brain is not that simple! Didn't somebody say that if the brain were so simple that we could understand it, then we would be so simple that we couldn't?"

"Be that as it may," they concede, "how does the idea of the quantum help with consciousness?"

I tell them about consciousness affecting the quantum wave. "Look, this is a paradox if consciousness is made of atoms. But if we flip our view of what the world is made of, this paradox is very satisfactorily resolved. I assure you, the world is made of consciousness." I can't conceal my excitement and even pride—it is such a powerful idea. I plead with them to join me.

"The sad thing," I continue, "is that if ordinary people really knew that consciousness and not matter is the link that connects us with each other and the world, then their views about war and peace, environmental pollution, social justice, religious values, and all other human endeavors would change radically."

"That sounds interesting, and I sympathize, believe me. But your idea also sounds like something out of a Good Book. How can we adopt religious ideas as science and still be credible?" The questioner sounds like he is talking to himself.

"I am asking you to give consciousness its due," I reply. "My friend Guernica needs consciousness to become whole again. And from what I've heard at this party, he's not the only one. How can you still debate whether consciousness even exists? Enough is enough. Surely the existence of consciousness is not debatable, and you know it."

"I see," says the fellow with the ponytail, shaking his head. "My friend, there's been a misunderstanding. We have all chosen to be Guernica; you have to if you want to do science. We have to assume that we are all made of atoms. Our consciousness has to be a secondary phenomenon—an epiphenomenon—of the dance of atoms. The essential objectivity of science demands it."

I go back to Guernica and sadly tell him my experience. "As Abraham Maslow once said, 'If the only tool you have is a hammer, then you start treating everything as if it were a nail.' These people are used to seeing the world as made of atoms and separate from themselves. They see consciousness as an illusory epiphenomenon. They can't grant you consciousness."

"But how about you?" Guernica gazes at me intensely. "Are you also going to hide behind scientific objectivity, or are you going to do something to help me regain wholeness?" He is shaking me now.

His intensity wakes me from the dream. Slowly, a resolve is born to write this book.

* * *

Today in physics, we face a great dilemma. In quantum physics—the new physics—we have found a theoretical framework that works; it explains myriad laboratory experiments and more. Quantum physics has led to such tremendously useful technologies as transistors, lasers, and superconductors. Yet we cannot make sense of the mathematics of quantum physics without suggesting an interpretation of experimental results that many people can only look upon as paradoxical, even impossible. Behold the following quantum properties:

- A quantum object (for example, an electron) can be at more than one place at the same time (*the wave property*).
- A quantum object cannot be said to manifest in ordinary space-time reality until we observe it as a particle (*collapse of the wave*).
- A quantum object ceases to exist here and simultaneously appears in existence over there; we cannot say it went through the intervening space (*the quantum jump*).
- A manifestation of one quantum object, caused by our observation, simultaneously influences its correlated twin object—no matter how far apart they are (*quantum action-at-a-distance*).

We cannot connect quantum physics with experimental data without using some schema of interpretation, and interpretation depends on the philosophy we bring to bear on the data. The philosophy that has dominated science for centuries (physical, or material, realism) assumes that only matter—consisting of atoms or, ultimately, elementary particles—is real; all else are secondary phenomena of matter, just a dance of the constituent atoms. This worldview is called realism because objects are assumed to be real and independent of subjects, us, or of how we observe them.

The notion, however, that all things are made of atoms is an unproven assumption; it is not based on any direct evidence for all things. When the new physics confronts us with a situation that seems paradoxical from the perspective of material realism, we tend to overlook the possibility that the paradoxes may be arising because of the falsity of our unproven assumption. (We tend to forget that a long-held assumption does not thereby become a fact, and we often even resent being reminded.)

Many physicists today suspect that something is wrong with material realism but are afraid to rock the boat that has served them so well for so long. They do not realize that their boat is drifting and needs new navigation under a new worldview.

Is there an alternative to the philosophy of material realism? Material realism strains unsuccessfully, notwithstanding its computer models, to explain the existence of our minds, especially the phenomenon of a causally potent self-consciousness, "What's consciousness?" The material realist tries to shrug away the question by answering cavalierly that it doesn't matter. If, however, we take all the theories that the conscious mind constructs (including the ones that negate it) with any seriousness, consciousness does matter.

Since René Descartes divided reality into two separate realms—mind and matter—many people have tried to rationalize the causal potency of conscious minds within Cartesian dualism. Science, nevertheless, presents compelling reasons to doubt that a dualistic philosophy is tenable: In order for the worlds of mind and matter to interact, they must exchange energy, yet we know that the energy of the material world remains constant. Surely, then, there is only one reality. Here is the catch 22: If the one reality is material reality, consciousness cannot exist except as an anomalous epiphenomenon.

So the question is, Is there a monistic alternative to material realism, where mind and matter are integrally part of one reality, but a reality that is not based on matter? I am convinced there is. The alternative that I propose in this book is monistic idealism. This philosophy is monistic as opposed to dualistic, and it is idealism because ideas (not to be confused with ideals) and the consciousness of them are considered to be the basic elements of reality; matter is considered to be secondary. In other words, instead of positing that everything (including consciousness) is made of matter, this philosophy posits that everything (including matter) exists in and is manipulated from consciousness. Note that the philosophy does not say that matter is unreal but that the reality of matter is secondary to that of consciousness, which itself is the ground of all being—including matter. In other words, in answer to What's matter? a monistic idealist would never say, Never mind.

This book shows that the philosophy of monistic idealism provides a paradox-free interpretation of quantum physics that is logical, coherent, and satisfying. Moreover, mental phenomena—such as self-consciousness, free will, creativity, even extrasensory perception—find simple, satisfying explanations when the mind-body problem is reformulated in an overall context of monistic idealism and quantum theory. This reformulated picture of the brain-mind enables us to understand our whole self entirely in harmony with what the great spiritual traditions have maintained for millennia.

The negative influence of material realism on the quality of modern human life has been staggering. Material realism poses a universe without any spiritual meaning: mechanical, empty, and lonely. For us—the inhabitants of the cosmos—this is perhaps the more unsettling because, to a frightening degree, conventional wisdom holds that material realism has prevailed over theologies that propose a spiritual component of reality in addition to the material one.

The facts prove otherwise; science proves the potency of a monistic philosophy over dualism—over spirit separated from matter. This book presents a strong case, supported by existing data, that the monistic philosophy needed now in the world is not materialism but idealism.

In the idealist philosophy, consciousness is fundamental; thus our spiritual experiences are acknowledged and validated as meaningful. This philosophy accommodates many of the interpretations of human spiritual experience that have sparked the various world religions. From this vantage point we see that some of the concepts of various religious traditions become as logical, elegant, and satisfying as the interpretation of experiments of quantum physics.

Know thyself. This has been the advice through the ages of philosophers who were quite aware that our self is what organizes the world and gives it meaning; to know the self along with nature was their comprehensive objective. Modern science's embracing of material realism changed all that; instead of being united with nature, consciousness became separate from nature, leading to a psychology separate from physics. As Morris Berman notes, this material realist worldview exiled us from the enchanted world in which we lived in yesteryear and condemned us to an alien world.[6] Now we live like exiles in this alien land; who but an exile would risk destroying this beautiful earth with nuclear war and environmental pollution? Feeling like exiles undermines our incentive to change our perspective. We are conditioned to believe that we are machines—that all our actions are determined by the stimuli we receive and by our prior conditioning. As exiles, we have no responsibility, no choice; our free will is a mirage.

This is why it has become so important for each of us to examine closely our worldview. Why am I being threatened by nuclear annihilation? Why does warfare continue as the barbaric way to settle the world's disputes? Why is there recurrent famine in Africa when we in the United States alone can grow enough food to feed the world? How did I acquire a worldview (more importantly, am I stuck with it?) that dictates so much separateness between me and my fellow humans, all of us sharing similar genetic, mental, and spiritual endowments? If I disown the outdated worldview that is based on material realism and investigate the new/old one that quantum physics seems to demand, might the world and I be once more integrated?

We need to know about us; we need to know if we can change our perspectives—if our mental makeup permits it. Can the new physics and the idealist philosophy of consciousness give us new contexts for change?

Chapter 2

THE OLD PHYSICS AND ITS PHILOSOPHICAL LEGACY

SEVERAL DECADES AGO, the American psychologist Abraham Maslow formulated the idea of a hierarchy of needs. After human beings satisfy their basic survival needs, it becomes possible for them to strive toward the fulfillment of higher-level needs. To Maslow the highest of these needs is the spiritual: the desire for self-actualization, for knowledge of oneself at the deepest possible level.[1] Since many Americans, in fact many Westerners, have already passed through the lower rungs of Maslow's ladder of needs, we should expect to see Westerners enthusiastically mounting the upper rungs toward self-actualization or spiritual fulfillment. We do not. What is wrong with Maslow's argument? As Mother Teresa observed when she visited the United States in the eighties, Americans are materially blessed but impoverished in spirit. Why should this be so?

Maslow neglected to take into account the consequences of unquestioned materialism, which is dominant in today's Western culture. Most Westerners accept as scientific fact the idea that we live in a materialist world—a world in which everything is made of matter and where matter is the fundamental reality. In such a world, material needs proliferate, resulting in desire not for spiritual progress but for more, bigger, and better things: bigger cars, better housing, the newest fashions, amazing forms of entertainment, and a dazzling extravaganza of present and future technological goodies. In such a world, our spiritual needs are often unrecognized, denied, or sublimated when they surface. If only matter is real, as materialism has taught us to believe, then material possessions are the only reasonable foundation for happiness and the good life.

Of course, our religions, our spiritual teachers, and our artistic and literary traditions teach that such is not the case. On the contrary, they teach that materialism leads, at best, to a sickening surfeit and, at worst, to crime, disease, and other ills.

Most Westerners hold both of these conflicting beliefs and live with ambivalence, partaking of a ravenously materialistic consumer culture yet secretly despising themselves for it. Those of us who still consider ourselves religious are not altogether able to ignore that, although our words and thoughts adhere to religion, all too often our deeds violate our intentions; we fail to embody with conviction even the most basic teachings of religions, such as kindness to our fellow humans. Others of us resolve our cognitive dissonance by embracing religious fundamentalism or equally fundamentalist scientism.

In sum, we live in a crisis—not so much a crisis of faith as a crisis of confusion. How did we reach this sorry state? By accepting materialism as the so-called scientific view of the world. Convinced that we must be scientific, we are like the keeper of an old curio shop in the following tale: A customer, finding an unfamiliar instrument, brought it to the shopkeeper and asked what it was for.

"Oh, that's a barometer," answered the shopkeeper. "It tells you if it's going to rain."

"How does it work?" the man wondered aloud.

The shopkeeper actually did not know how a barometer works but to admit that would be to risk losing a sale. So the shopkeeper said, "You hold it out the window and then bring it back in. If the barometer is wet, you know it is raining."

"But I could do that with my bare hand, so why use a barometer?" the man protested.

"That would not be scientific, my friend," responded the shopkeeper.

I submit that in our acceptance of materialism, we are like the shopkeeper. We want to be scientific; we think we are being scientific, but we are not. To be truly scientific, we must remember that science has always changed as it has made new discoveries. Is materialism the correct scientific worldview? I believe that the answer is demonstrably no, although scientists themselves are confused on this issue.

The scientist's confusion is due to a hangover caused by an overly enthusiastic indulgence in a four-hundred-year-old revel called classical physics that was kicked off by Isaac Newton sometime around 1665. Newton's theories launched us on the course that led to the materialism that dominates Western culture. The philosophy of materialism, which dates back to the Greek philosopher Democritus (ca. 460–ca. 370 B.C.), matches the worldview of classical physics which is variously termed material, physical, or scientific realism. Although a new scientific discipline called quantum physics has formally replaced classical physics in this century, the old philosophy of classical physics—that of material realism—is still widely accepted.

Classical Physics and Material Realism

When he visited the palace at Versailles, the seventeenth-century French mathematician and philosopher René Descartes was enchanted by the huge assembly of automata in the palace garden. Driven by unseen mechanisms, water flowed, music played, sea nymphs frolicked, and mighty Neptune rose from under a pool. As he watched the display, Descartes conceived the idea that the world might be such an automaton—a world machine.

Descartes later propounded a significantly modified version of his picture of the world as a machine. His famous philosophy of dualism divided the world into an objective sphere of matter (the domain of science) and a subjective sphere of mind (the domain of religion). Thus did Descartes free scientific investigation from the orthodoxy of the powerful church. Descartes borrowed the idea of objectivity from Aristotle. The basic notion is that objects are independent of and separate from the mind (or consciousness). We will refer to this as the principle of *strong objectivity*.

Descartes also made contributions to the laws of physics that would scientifically enshrine his idea of the world as a machine. It was, however, Newton and his heirs going into the eighteenth century who solidly established materialism and its corollary: the principle of *causal determinism*, which is the idea that all motion can be predicted exactly given the laws of motion and the initial conditions on the objects (where they are and with what velocity they are moving).

To understand the Cartesian-Newtonian view of the world, think of the universe as a big bunch of billiard balls—large and small—in a three-dimensional billiard table that we call space. If we know all the forces acting on each of these billiard balls at all times, then just knowing their initial conditions—their positions and velocities at some initial time—enables us to calculate where each of these bodies will be at all future times (or, for that matter, where they were at any previous time).

The philosophical import of determinism was best summarized by the eighteenth-century mathematician Pierre-Simon de Laplace: "An intelligence that, at a given instant, was acquainted with all the forces by which nature is animated and with the state of the bodies of which it is composed, would—if it were vast enough to submit the data to analysis—embrace in the same formula the movements of the largest bodies of the universe and those of the lightest atoms: nothing would be uncertain to such an intelligence, and the future, like the past, would be present to its eyes."[2]

Laplace also wrote a successful book on celestial mechanics that made him famous, so famous that the emperor Napoleon summoned him to the palace.

"Monsieur Laplace," said Napoleon, "you have not mentioned God in your book even once. Why is that?" (In those days, custom demanded that God be cited a few times in any book of consequence, so Napoleon was obviously curious. What kind of daring individual was this Laplace to break with such a venerable custom?) Laplace's purported reply is a classic:

"Your majesty, I have not needed that particular hypothesis."

Laplace properly understood the implication of classical physics and its causally deterministic mathematical framework. In a Newtonian universe, God is not needed!

We have now learned two fundamental principles of classical physics: strong objectivity and determinism. A third principle of classical physics was discovered by Albert Einstein. Einstein's theory of relativity, an extension of classical physics to bodies that move with high velocity, demanded that the highest velocity on nature's highways be the velocity of light. This velocity is enormous—300,000 kilometers per second—but even so it is limited. The implication of this speed limit is that all influences between material objects happening in space-time must be local: They must travel through space one bit at a time with a finite velocity. This is called the principle of *locality*.

When Descartes divided the world into matter and mind, he intended a tacit agreement not to attack religion, which would reign supreme in matters of the mind, in exchange for science's supremacy over matter. For more than two hundred years, the agreement held. Eventually the success of science in predicting and controlling the environment prompted scientists to question the validity of any religious teaching. In particular, scientists began to challenge the mind, or spirit, side of Cartesian dualism. The principle of *material monism* was thus added to the list of postulates of material realism: All things in the world, including mind and consciousness, are made of matter (and of such generalizations of matter as energy and force fields). Ours is a material world through and through.

Of course, nobody yet knows how to derive mind and consciousness from matter, thus another promissory postulate was added: the principle of *epiphenomenalism*. According to this principle, all mental phenomena can be explained as epiphenomena, or secondary phenomena, of matter by a suitable reduction to antecedent physical conditions. The basic idea is that what we call consciousness is simply a property (or group of properties) of the brain when the brain is viewed at a certain level.

These five principles, then, constitute the philosophy of material realism:

1. Strong objectivity
2. Causal determinism
3. Locality
4. Physical or material monism
5. Epiphenomenalism

✓ pg 46

This philosophy is also called scientific realism, which implies that material realism is essential to science. Most scientists, at least unconsciously, still believe that this is so, even in the face of firmly established data that contradict the five principles.

It is important to realize at the outset that the principles of material realism are metaphysical postulates. They are assumptions about the nature of being, not conclusions arrived at by experiment. If experimental data are discovered that contradict any of these postulates, then that postulate must be sacrificed.

Similarly, if rational argumentation reveals the weakness of a particular postulate, the validity of that postulate must be questioned.

A major weakness of material realism is that the philosophy seems to exclude subjective phenomena altogether. If we hold on to the postulate of strong objectivity, many powerful experiments done in the cognitive laboratory will not be admissible as data. Material realists are quite aware of this shortcoming; thus in recent years much attention has been given to the question of whether mental phenomena (including self-consciousness) can be understood on the basis of material models—notably, computer models. We shall examine the basic idea behind such models: the idea of the mind machine.

CAN WE BUILD A COMPUTER THAT IS CONSCIOUS?

The challenge for science after Newton was, of course, to attempt to approximate as closely as possible Laplace's all-knowing intelligence. The insight of Newtonian classical physics proved to be quite powerful, and significant strides were made toward such an approximation. Scientists gradually unraveled, at least in part, some of the so-called eternal mysteries—how our planet came into being, how stars find their energy to burn, how the universe was created, and how life reproduces itself.

Eventually, the successors of Laplace took on the challenge of explaining the human mind, self-consciousness and all. With their deterministic insight, they had no doubt that the human mind also was a Newtonian classical machine, like the world machine of which it was a part.

One of the believers in mind-as-machine, Ivan Pavlov, was very gratified with his dogs' confirmation of his belief. When Pavlov rang a bell, his dogs salivated, even though no food was offered. The dogs had been conditioned to expect food whenever the bell rang, explained Pavlov. It was quite simple, really. Give a stimulus, observe the response, and if it is the one you want, reinforce it with a reward.

Thus was born the idea that the human mind was a simple machine, with simple input-output statements in one-to-one correspondence that operate on a stimulus-response-reinforcement basis. This idea was much criticized on the ground that such a simple behavioral machine could not carry out such mental processes as thinking.

You want thinking, you've got it, replied the clever classical machinists, who conceived the idea of a complex machine with internal states. Look at the behavior of even a simple mobile, they said. It is such fun to watch because its responses to a wind pattern are infinitely varied. Why? Because each response literally depends on the many juxtapositions of the various internal states of the branches of the mobile, in addition to the specific stimulus. For the brain, these internal states are synonymous with thinking, feeling, and so forth, which are the epiphenomena of the internal states of the complex machine that is the human brain.

The voices of opposition still protested: What about free will? Human beings have freedom of choice. The machinists responded that free will is but an illusion; they added the interesting argument that there is a possible physical model of the illusive free will. The ingenuity of the researchers of the mind-machines is truly admirable. There is now the idea that, although classical systems are ultimately deterministic, displaying a basically deterministic behavior, we can also have chaos: On occasion, very slight changes in initial conditions can produce very large differences in the final outcome for a system.[3] This generates uncertainty (the uncertainty of weather systems is an example of this chaotic behavior), and uncertainty of prediction can be interpreted as free will. Because chaos is ultimately determined chaos, so the argument goes, this is an illusion of free will. So, is our free will an illusion?

An even more convincing argument in favor of the machine picture of ourselves has come from the British mathematician Alan Turing. Someday, says Turing, we shall construct a machine that follows classical deterministic laws—a silicon computer that will carry on a conversation with any of us humans who have so-called free will. He challenges, moreover, that impartial observers will not be able to discern the conversation of the computer from that of the human being.[4] (I propose that this be the credo of a new society: OEHAI, the Organization for the Equality of Human and Artificial Intelligence.)

Although I admire much of the progress in the area of artificial intelligence, I am unconvinced that my consciousness is an epiphenomenon and that my free will is a mirage. I do not recognize the limits that locality and causality impose on a classical machine as my limits. I do not believe that these are real limits for any human being, and I am concerned that thinking they are may be a self-fulfilling prophecy.

"We are the mirrors of the world in which we dwell," said the science historian Charles Singer. The question is, How big a mirror can we be? Reflections of the sky are found in little ponds and in the mighty ocean. Which is the bigger reflection?

But we have come a long way toward developing an intelligent Turing machine, protest the mind-machine proponents. Our machines already can pass the Turing test with an occasional unsuspecting human. Surely, with further nurturing and development they will have minds like those of humans. They will understand, learn, and behave like us.

If we can make Turing machines that behave like humans in every known way, the mind-machinist continues in a determined voice, isn't that proof that our own minds are nothing but a bunch of classical computer programs, utterly determined? Since determined is not the same as predictable, the unpredictability of humans presents no obstacle to the view. This argument is persuasive as far as it goes. If our computers can simulate human behavior, good; this will make communication easier between us and our machines. If by studying the workings of the computer programs that simulate some of our behavior, we learn something about ourselves, that is even better. Simulating our behavior on computers, however, is a long way from proving that we are made of those programs that do the simulations.

Of course, even one example of a program we possess that a classical computer can never duplicate will destroy the myth of the mind-as-machine. The mathematician Roger Penrose argues that computerlike, algorithmic reasoning is insufficient for the discovery of mathematical theorems and laws. (An algorithm is a systematic procedure for solving a problem: a strictly logical, rule-based approach.) So, asks Penrose, where does mathematics come from if we operate like a computer? "Mathematical truth is *not* something that we ascertain merely by use of an algorithm. I believe, also, that our *consciousness* is a crucial ingredient in our comprehension of mathematical truth. We must 'see' the truth of a mathematical argument to be convinced of its validity. This 'seeing' is the very essence of consciousness. It must be present *whenever* we directly perceive mathematical truth."[5] In other words, our consciousness must exist prior to our algorithmic computer capacity.

An even stronger argument against the position of mind-as-machine is pointed out by the Nobel laureate physicist Richard Feynman.[6] A classical computer, notes Feynman, can never simulate nonlocality (a technical word meaning information or influence transfer without local signals; such influences are action-at-a-distance and instantaneous). Thus, if nonlocal information processing exists in humans, it is one of our nonalgorithmic programs that a classical computer can never simulate.

Do we have nonlocal information processing? We can make a very good case for nonlocality if we accept our spirituality. Another controversial case for nonlocality is the claim of paranormal experiences. People through the centuries have claimed the capacity for telepathy, mind-to-mind transmission of information without local signals, and now there seems to be some scientific evidence for it.[7]

Alan Turing himself realized that telepathy is one sure way for an interrogator to discern a human being from a silicon computing machine in a Turing test: "Let us play the imitation game, using as witnesses a man who is good as a telepathic receiver and a digital computer. The interrogator can ask such questions as 'What suit does the card in my hand belong to?' The man by telepathy or clairvoyance gives the right answer 130 times out of 400 cards. The machine can only guess at random, and perhaps gets 104 right, and so the interrogator makes the right identification."[8]

Extrasensory perception (ESP), which admittedly remains controversial, is only one case against the power of the classical computer. Another important capability of the human mind that seems beyond the scope of the silicon computer is creativity. If creativity involves discontinuity, abrupt breaks from past patterns of thinking, then the computer's ability to be creative is certainly suspect because a classical computer operates with continuity.[9]

Ultimately, however, the crux of the matter is consciousness. If the mind-machinists can develop a classical computer that is conscious in the sense that you and I are, it will become a different ballgame in spite of all of the above circumstantial considerations. Can they? How could we tell? Suppose we equipped a Turing machine with zillions of programs that simulated our own behavior perfectly; would the machine then become conscious? Surely, her (assuming the machine were built to be female) behavior would show all the complexities of the human mind, and as a Turing machine, she would be an impeccable simulation of a human (except for a few such distinctively human traits as ESP and mathematical creativity, which the mind-machinists regard as dubious, anyway), but would she be truly conscious?

When I was in college, in the fifties, I became aware of the idea of a conscious computer while reading a science fiction novel by Robert Heinlein: *The Moon Is a Harsh Mistress.* Heinlein conveyed the notion that computer consciousness is a question of the computer's size and complexity; as soon as the machine in the novel passed a threshold of size and complexity, it became conscious. This view seems to have wide currency among the many researchers who play the computer-mind game.

I feel that the question of computer consciousness is not a question of complexity. Admittedly, a high level of complexity can guarantee that the responses of a computer under a given stimulus will be no more easily predictable than are a human's, but it means no more than that. If we can trace the computer's input-output performance to the activities of its internal circuits without any ambiguity, without losing the trail (and this, at least in principle, should always be possible for a classical computer), then what is the necessity for consciousness? It would seem to have no function. I think it is an evasion of the issue for artificial intelligence protagonists to say that consciousness is only an epiphenomenon, or an illusion. The Nobel laureate neurophysiologist John Eccles seems to agree with me. Asks Eccles: "Why do we have to be conscious at all? We can, in principle, explain all our input-output performances in terms of the activity of the neuronal circuits; and consequently consciousness seems to be absolutely unnecessary."[10]

Not everything that is unnecessary is forbidden in nature, but it is not likely to occur. Consciousness seems unnecessary for a classical Turing machine, and this is reason enough to doubt that these machines, however sophisticated, will ever be conscious. The fact that we do have consciousness suggests only that our input-output performances are not wholly determined by the algorithmic programs of classical computer machinery.

The mind-machinists sometimes pose another argument: We freely assign consciousness to other human beings because they report mental experiences—thoughts, feelings—that are similar to our own. If an android were programmed to report thoughts and feelings similar to yours, could you discern its consciousness from that of your friend? After all, you cannot experience what is inside your human friend's head any more than you can experience what is inside the android's. Thus you can never really know, anyway!

This reminds me of an episode of the television show "Star Trek." A con man is given an unusual punishment that on the face of it seems to be no punishment at all. He is banished to a colony where he will be the only human, surrounded by androids at his service—many of them in the form of beautiful maidens.

You can guess as well as I can why this is a punishment. The reason that I do not live in a solipsistic (only I am real) universe is not that others like me logically convince me of their humanness but that I have an inner connection with them. I could never have this connection with an android.

I submit that the sense we have of an inner connection with other humans is due to a real connection of the spirit. I believe that classical computers can never be conscious like us because they lack this spiritual connection.

Etymologically, the word *consciousness* derives from the words *scire* (to know) and *cum* (with). Consciousness is "to know with." To me, this term implies nonlocal knowing; we cannot know with somebody without sharing a nonlocal connection with that person.

It should cause no dismay if we cannot build a model of ourselves based on classical physics and using a silicon computer's algorithmic approach. We have known from the beginning of this century that classical physics is incomplete physics. No wonder it gives us an incomplete worldview. Let us examine the new physics, born at the dawn of the twentieth century, and explore, from our vantage point as the century draws to a close, what freedom its worldview brings.

Chapter 3

QUANTUM PHYSICS AND THE DEMISE OF MATERIAL REALISM

ALMOST A CENTURY AGO, a series of experimental discoveries was made in physics that called for a change in our worldview. What started showing up were, in the words of philosopher Thomas Kuhn, anomalies that could not be explained by classical physics.[1] These anomalies opened the door to a revolution in scientific thought.

Imagine that you are a physicist at the turn of the century. One of the anomalies you and your colleagues are interested in understanding is how hot bodies emit radiation. As a physicist of Newtonian vintage, you believe that the universe is a classical machine consisting of parts that behave according to Newtonian laws that are almost all completely known. You believe that once you have all the information about the parts and have figured out the few remaining glitches about the laws, you will be able to predict the future of the universe forever. Still, those few glitches are troubling. You are not prepared to answer such questions as, What is the law of emission of radiation from hot bodies?

Imagine, as you puzzle over the question, that your loved one is comfortably seated beside you in front of a glowing fire.

YOU (*muttering*): I just can't figure this out.
LOVED ONE: Pass the nuts.

YOU (*while passing the nuts*): I just can't figure out why we are not getting a good tan right now.

LO (*laughing*): Well, that would be nice. We could even justify using the fireplace in the summertime.

YOU: You see, theory says that the radiation from the fireplace should be as rich in high-frequency ultraviolet as sunlight is. But what makes sunlight and not fireplace light rich in these high frequencies? Why aren't we tanning in an ultraviolet bath right now?

LO: Wait a minute, please. If I am going to listen to this seriously, you'll have to slow down a little and explain. What's frequency? What's ultraviolet?

YOU: Sorry. Frequency is the number of cycles per second. It's the measure of how fast a wave wiggles. For light, that means color. White light is made up of light of various frequencies, or colors. Red is low-frequency light, and violet is high-frequency light. If the frequency is even higher, it's invisible black light, what we call ultraviolet.

LO: Okay, so light from both burning wood and the sun should give out plenty of ultraviolet. Unfortunately, the sun follows your theory, but burning wood doesn't. Maybe there's something special about burning wood. . .

YOU: Actually, it's even worse than that. All light sources, not just the sun or burning wood, should give off copious amounts of ultraviolet.

LO: Ah, the plot thickens. The inflation of the ultraviolet is ubiquitous. But isn't all inflation followed by a recession? Isn't there a song, what goes up must come down? (*Your loved one starts humming.*)

YOU (*exasperated*): But how?

LO (*holding out the bowl of nuts*): Nuts, dear?

(The conversation ends.)

PLANCK TAKES THE FIRST QUANTUM JUMP

Many physicists in the late nineteenth century were frustrated until, finally, one of them broke rank: Max Planck, of Germany. In 1900, Planck took a bold conceptual leap and said that what the old theory needed was a quantum jump. (He borrowed the word *quantum*, meaning "amount," from Latin.) What emitted the light from an incandescent body—burning wood, for example, or the Sun—were tiny jiggling charges, the electrons. These electrons absorb energy from a hot environment, such as a fireplace, and then emit it back as radiation. This part of the old physics was correct, but then classical physics predicts that the emitted radiation should be rich in ultraviolet, which is contradicted by our observations. Planck declared (very bravely) that if the electrons are assumed to emit or absorb energy only in certain specific, discontinuously discrete amounts—which he called "quanta" of energy—the problem of the emission of varying degrees of ultraviolet could be solved.

To explore the meaning of the quantum of energy, consider an analogy. Compare the case of a ball on a staircase with one on a ramp (fig. 1). The ball on the ramp can assume any position, and its position can change by any amount. It is, therefore, a model of continuity and represents how we think in classical physics. In contrast, the ball on the staircase can sit only on this step or that; its position (and its energy, which is related to position) is "quantized."

You may object. What happens when the ball falls from one step to another? Is it not taking on an intermediate position during the descent? This is where the strangeness of quantum theory enters: For a ball on a set of stairs, the answer is obviously yes, but for a quantum ball (an atom or an electron), Planck's theory answers no. A quantum ball will never be found in any place intermediate between two steps; it is either on this one or on that one. This is a quantum discontinuity.

So why can you not get a tan from a wood-burning fireplace? Imagine a pendulum in the wind. Ordinarily a pendulum will swing in such a situation, even when there is not a high wind. Suppose, however, that the pendulum is allowed to absorb energy only in discrete steps of high denominations. In other words, it is a quantum pendulum. What then? Clearly, unless the wind is able to impart the required high increment of energy in one step, the pendulum will not move. Accepting energy in small denominations will not enable it to build up enough energy to cross a threshold. So it is with the jiggling electrons in a fireplace. Low-frequency radiation arises from small quantum jumps, but high-frequency radiation requires large quantum jumps. A large quantum jump must be fueled by a large amount of energy in the electron's environment; the energy in a wood-burning fireplace simply is not strong enough

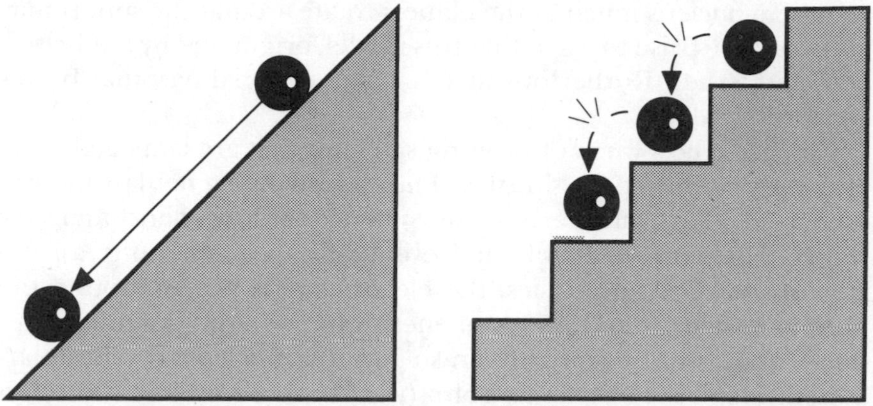

Figure 1. The quantum jump. On the ramp, the classical motion of the ball is continuous; on the staircase, quantum motion acts in discontinuous steps (quantum jumps).

to create the conditions even for much blue light, let alone ultraviolet. That is the reason you cannot get a tan from a fireplace.

By all accounts, Planck was a rather traditional sort of guy and declared his ideas about quanta of energy reluctantly. He even used to work on his mathematics standing up, as was at that time customary in Germany. He did not particularly like the implications of his breakthrough idea; that it pointed to a whole new way of understanding our physical reality was becoming clear, however, to scientists who would carry the revolution much further.

Einstein's Photons and Bohr's Atom

One of these revolutionaries was Einstein. He was working as a clerk at the patent office in Zurich at the time he published his first research paper on the quantum theory (1905). Challenging the then-popular belief that light is a wave phenomenon, Einstein suggested that light exists as a quantum—a discrete bundle of energy—that we now call a photon. The higher the frequency of the light, the more energy in each bundle.

Even more revolutionary was Danish physicist Niels Bohr, who in 1913 applied the idea of light quanta to suggest that the whole world of the atom is full of quantum jumps. We all have been taught that the atom resembles a tiny solar system, that electrons rotate around a nucleus much as the planets rotate around the sun. It may come as a surprise to learn that this model, originated by the British physicist Ernest Rutherford in 1911, has a crucial flaw that Bohr's work resolved.

Consider the swarm of orbiting satellites that are launched quite regularly by our space shuttles. These satellites do not last forever. Due to collisions with Earth's atmosphere, they lose energy and slow down. Their orbits shrink, and eventually they crash (fig. 2). According to classical physics, the electrons that swarm around the atomic nucleus would also lose energy, by radiating light continuously, and would eventually crash into the nucleus. So the solar-system atom is not stable. Bohr (who supposedly saw the solar-system atom in a dream), however, created a stable model of the atom by applying the concept of the quantum jump.

Suppose, said Bohr, that the orbits that electrons describe are discrete, like the quanta of energy suggested by Planck. The orbits

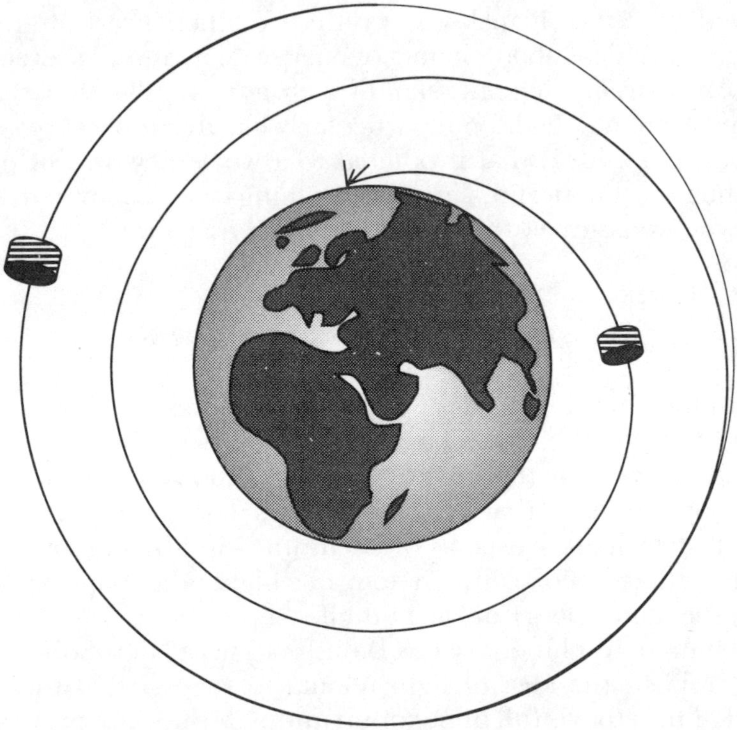

Figure 2. The orbits of satellites around the earth are unstable. The orbits of the electrons in the Rutherford atom behave in the same way.

can then be looked upon as making up an energy staircase (fig. 3). They are stationary—nonchanging in their energy value. The electrons, while in these quantized stationary orbits, do not radiate light. Only when an electron jumps from a higher-energy orbit to one of lower energy (from a higher level of the energy staircase to a lower level) does it emit light as a quantum. Thus, if an electron is in its lowest-energy orbit, it has no lower level to which it can jump. This ground-level configuration is stable, and there is no chance of an electron crashing into the nucleus. Physicists everywhere greeted Bohr's model of the atom with a sigh of relief.

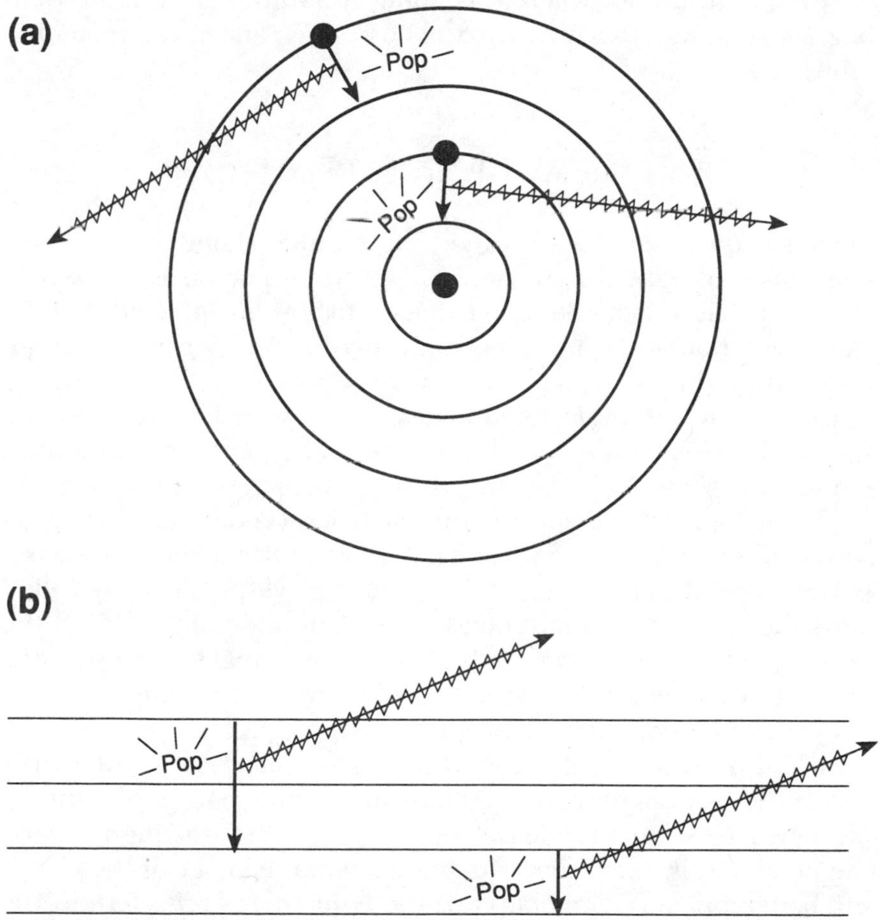

Figure 3. Bohr orbit and the quantum jump. (a) Bohr's quantized orbits. Atoms emit light when electrons jump orbits. (b) To quantum jump the energy ladder, you do not have to go through the intervening space between rungs.

Bohr had cut off the Hydra's head of instability, but another grew in its place. The electron, according to Bohr, can never occupy any position between orbits; thus when it jumps, it must somehow transfer directly to another orbit. This is not an ordinary jump through space but something radically new. Although you might be tempted to picture the electron's jump as a jump from one rung of a ladder to another, the electron makes the jump without ever passing through the space between the rungs. Instead, it seems to disappear at one rung and to reappear at the other—quite discontinuously. There is more: We cannot tell when a particular electron is going to jump nor where it is going to jump if it has more than one lower rung from which to choose. We can only give probabilities.

THE WAVE-PARTICLE DUALITY

Perhaps you have noticed something strange about the quantum conception of light. To say that light exists as quanta, as photons, is to say that light is made of particles—like grains of sand. Such a statement, however, contradicts many ordinary experiences that we have with light.

Imagine, for example, looking at a distant streetlight through the fabric of a cloth umbrella. You will not see a continuous, uninterrupted stream of light pouring through, which is what you would expect if light were made of tiny particles. (Pour sand through a sieve and you will see what I mean.) Instead, what you will see is a pattern of alternating bright and dark fringes, technically called a diffraction pattern. Light bends in and around the threads of the fabric and creates patterns that only waves make. So even our ordinary experience shows that light behaves like a wave.

Quantum theory nevertheless insists that light also behaves like a bunch of particles, or photons. Our eyes are such wonderful instruments that we can observe the quantum, grainy nature of light for ourselves. Next time you leave your loved one in the twilight, watch the person walk away from you. Notice how the image of the receding body appears fragmentary. If the light energy reflected off the body and onto the optical receptors of your retina had a wavelike continuity, at least some light from every part of the body would always be exciting your optical receptors: You would always see a complete image. (Granted, in dim light the contrast between light and dark would not be very clear, but this would not affect the sharpness of the outline.) What you see instead, however, is not a sharp outline because the receptors of your eyes respond to individual photons. Dim light has fewer photons than does bright light; so in this hypothetical twilight scenario only a few of your receptors will be stimulated at any given time, too few to define the outline or shape of a dimly lit body. Consequently, the image that you see will be fragmentary.

One more question may be nagging you. Why can the receptors not store their data indefinitely until the brain has enough information to collect all the fragmentary pictures into one whole? Fortunately for the quantum physicist, who is always desperately in need of everyday examples of quantum phenomena, the optical receptors can store information for only a tiny fraction of a second. In dim light not enough receptors in your eyes will fire at any given time to create a complete image. When next you wave adieu to the misty, departing figure of your loved one in the twilight, don't forget to ponder the quantum nature of light; it will surely lessen the pain of your separation.

When light is seen as a wave, it seems capable of being in two (or more) places at the same time, as when it passes through the slits of an umbrella and produces a diffraction pattern; when we catch it on a photographic film, however, it shows up discretely, spot by spot, like a beam of particles. So light must be both a wave and a particle. Paradoxical, isn't it? At stake is one of the bulwarks of the old physics: unambiguous description in language. Also at stake is the idea of objectivity: Does the nature of light—what light is—depend on how we observe it?

As if these paradoxes regarding light were not provocative enough, another question inevitably arises: Can a material object, such as an electron, be both a wave and a particle? Can it have a duality like that of light? The physicist who first asked this question and doggedly suggested a profession-shaking answer in the affirmative was a prince in the French aristocracy, Louis-Victor de Broglie.

Matter Waves

When de Broglie was writing his Ph.D. thesis around the year 1924, he made an association between the discreteness of the stationary orbits of the Bohr atom and that of sound waves produced by a guitar. The connection is a fruitful one.

Imagine a wave of sound traveling through a medium (fig. 4). The vertical displacement of the particles of the medium fluctuates from zero to a maximum (crest), back to zero, to a negative maximum (trough), and back to zero, over and over again, as the distance increases. The maximum vertical displacement in one direction (crest, or trough, to zero) is called the amplitude. The individual particles of the medium move back and forth about their undisturbed position. The wave going through the medium, however, propagates: A wave is a propagating disturbance. The number of crests passing a given point in a second is called the frequency of the wave. The crest-to-crest distance is the wavelength.

Plucking a guitar string sets it in motion, but the resulting vibrations are called stationary because they do not travel beyond the string. At any given place on the string, the displacement of the particles of the string changes with time: There is waviness, but the waves do not propagate in space (fig. 5). The propagating waves that we hear are those that have been set in motion by the stationary waves of the vibrating strings.

A musical note from a guitar consists of a whole series of sounds—a spectrum of frequencies. The interesting thing for de Broglie was that the stationary waves along the guitar string make up a discrete frequency spectrum called the harmonics. The lowest-frequency sound is called the first harmonic, which determines the pitch we hear. The higher harmonics—the musical sounds in the note that give it a distinctive quality—have frequencies that are represented as integer multiples of that of the first harmonic.

Being stationary is a property of waves in confinement. Such

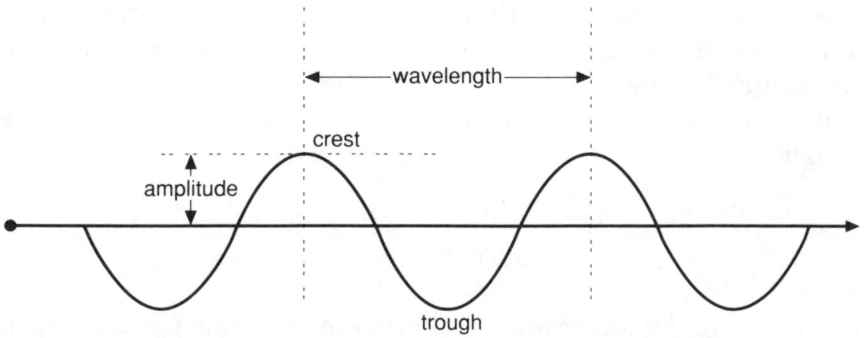

Figure 4. Graphic representation of a wave.

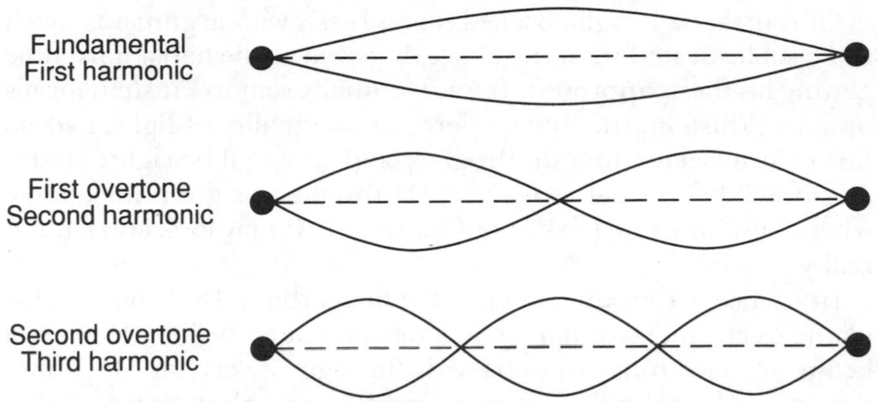

Figure 5. The first few harmonics of a standing or stationary wave on a guitar string.

waves are easily set up in a cup of tea. De Broglie asked, Are atomic electrons confined waves? If so, do they produce discrete stationary wave patterns? For example, maybe the lowest atomic orbit is one in which one electron makes a stationary wave of the smallest frequency—the first harmonic—and the higher orbits correspond to stationary electron waves of higher harmonics (fig. 6).

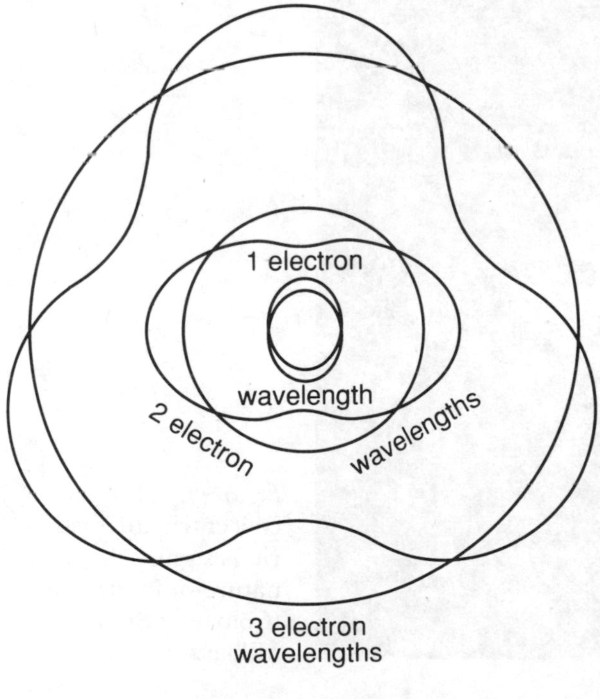

Figure 6. de Broglie's vision: Could electrons be stationary waves in the confinement of the atom?

Of course, de Broglie backed up his thesis with arguments much more sophisticated than the above, but even so, he had a hard time getting his thesis approved. It was eventually sent to Einstein for his opinion. Einstein, the first to perceive the duality of light, had no difficulty in seeing that de Broglie could very well be right: Matter might well be as dual as is light. De Broglie was given his degree when Einstein wrote back about his thesis: "It may look crazy, but it really is sound."

In science, experimentation is the final arbiter. De Broglie's idea of the electron's wave nature was demonstrated brilliantly when a beam of electrons was passed through a crystal (a three-dimensional "umbrella" suitable for diffracting electrons) and photographed. The result was a diffraction pattern (fig. 7).

If matter is a wave, quipped one physicist to another at the end of a seminar in 1926 on de Broglie's waves, there should be a wave equation to describe a matter wave. The physicist who said this promptly forgot about it, but the one who heard it, Erwin Schrö-

Figure 7. The concentric diffraction rings signify the wave nature of electrons. (Courtesy: Stan Miklavzina.)

dinger, proceeded to discover the wave equation for matter, now known as the Schrödinger equation. It is the cornerstone of the mathematics that replaced Newton's laws in the new physics. The Schrödinger equation is used to predict all the wonderful properties of submicroscopic objects that our laboratory experiments reveal. Werner Heisenberg had discovered the same equation even earlier but in a more obscure mathematical form. The mathematical formalism that grew out of the work of Schrödinger and Heisenberg is called quantum mechanics.

De Broglie's and Schrödinger's idea of the matter wave generates a remarkable picture of the atom. It explains in simple terms the three most important properties of atoms: their stability, their identity with one another, and their ability to regenerate themselves. I have already explained how stability arises—that was the great contribution of Bohr. The identity of atoms of a particular species is simply a consequence of the identity of wave patterns in confinement; the structure of the stationary patterns is determined by the manner in which the electrons are confined, not by their environment. The music of the atom, its wave pattern, is the same wherever you find it—on Earth or Andromeda. Furthermore, the stationary pattern, depending only on the conditions of its confinement, has no trace of past history, no memory; it regenerates itself, repeating the same performance over and over.

Probability Waves

Electron waves are no ordinary waves. Even in a diffraction experiment, the individual electrons show up at the photographic plate as localized individual events; only when we observe the pattern created by a whole bunch of electrons do we find evidence of their wave nature—the diffraction pattern. Electron waves are probability waves, said the physicist Max Born. They tell us probabilities: For example, where we are most likely to find the particle is where the wave disturbances (or the amplitudes) are strong. If the probability of finding the particle is small, the wave amplitude will be weak.

Imagine that you are watching traffic from a helicopter above the streets of Los Angeles. If the cars were described by Schrödinger's waves, we would say that the wave is strong at the location of traffic jams and that between jams the wave is weak.

Furthermore, electron waves are conceived of as *wave packets*. By employing the notion of packets we can make the wave amplitude large in specific regions of space and small everywhere else (fig. 8). This is important because the wave has to represent a localized particle. The wave packet is a packet of probability, and Born said that for electron waves, the square of the wave amplitude—technically called the wave function—at a point of space gives us the probability of finding the electron at that point. This probability can be represented as a bell-shaped curve (fig. 9).

THE HEISENBERG UNCERTAINTY PRINCIPLE

Probability begets uncertainty. For an electron or any other quantum object, we can speak only of the probability of finding the object at such and such a position or of its momentum (mass times velocity) being so and so, but these probabilities form a distribution such as that represented by the bell-shaped curve. The probability will be maximum for some value of the position, and this will be the most likely place to find the electron. But there will be an entire region of places where there will be a considerable chance of locating the electron. The width of this region represents the degree of uncertainty of the electron's position. The same argument enables us to talk about the uncertainty of the electron's momentum.

From such considerations, Heisenberg mathematically proved that the product of the uncertainties of the position and the momentum is greater than or equal to a certain small number called Planck's constant. This number, originally discovered by Planck, sets the quantitative scale at which quantum effects become appreciably large. If Planck's constant were not small, the effects of the quantum uncertainty would invade even our ordinary macro reality.

Figure 8. Superposition of many simple waves produces a typical localized wave packet. (Adapted with permission from P. W. Atkins, *Quanta: A Handbook of Concepts*. Oxford: Clarendon Press, 1974.)

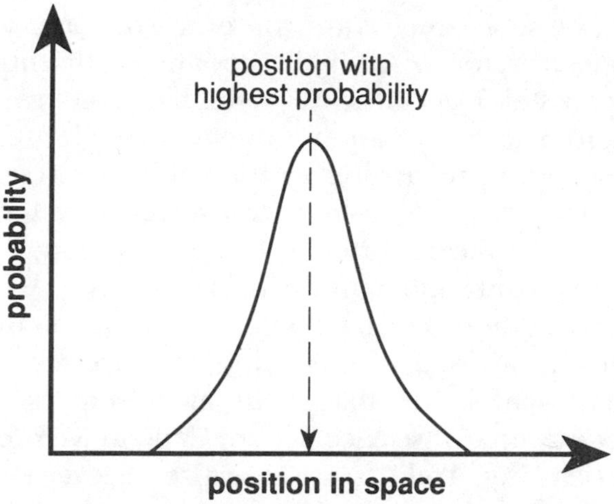

Figure 9. A typical probability distribution.

In classical physics, all motion is determined by the forces that govern it. Once we know the initial conditions (the position and the velocity of an object at some initial instant of time), we can calculate its precise trajectory using Newton's equations of motion. Thus classical physics leads to the philosophy of determinism, the idea that it is possible to predict completely the motion of all material objects.

The uncertainty principle throws a Molotov cocktail into the philosophy of determinism. According to the uncertainty principle, we cannot simultaneously determine with certainty both the position and the velocity (or momentum) of an electron; any effort to measure one accurately blurs our knowledge of the other. Thus the initial conditions for the calculation of a particle's trajectory can never be determined with accuracy, and the concept of a sharply defined trajectory of a particle is untenable.

By the same token, the Bohr orbits do not provide a strict description for the whereabouts of the electron: The position of the actual orbits is fuzzy. We really cannot say that the electron is such and such distance away from the nucleus when it is at this or that energy level.

Uncertain Fantasies

Consider a few fantasy scenarios in which the writers were unaware of or forgot the importance of the uncertainty principle.

In *Fantastic Voyage*, a science fiction book and movie, objects were miniaturized by squeezing. Have you ever wondered whether it is possible to squeeze atoms? After all, they are mostly empty space. Is such a thing possible? Decide for yourself by considering the uncertainty relation. The size of an atom gives a rough estimate of the degree of uncertainty regarding position of its electrons. Squeezing the atom will localize its electrons within a smaller volume of space, thus reducing the uncertainty about their position; but then the uncertainty regarding momentum must increase. An increase in the uncertainty of the electron's momentum means an increase in its velocity. Thus, as a result of squeezing, the electrons' velocity increases, and they are better able to run away from the atom.

In another example of science fiction, Captain Kirk (of the classic television series "Star Trek") says, Energize. A lever is then pulled down on an instrument panel; voilà, people standing on a platform disappear and reappear at a destination that is supposed to be an unexplored planet but that looks a lot like a Hollywood sound stage. In one of his novels based on "Star Trek," James Blish tried to characterize this process of reappearing as a quantum jump. Just as an electron jumps from one atomic orbit to another without ever passing through the intermediate space, so would the crew of the spaceship *Enterprise*. You can see a problem with this. When the electron will take the jump and to where are acausal and unpredictable because probability and uncertainty rule the quantum jump. Such quantum transport would force the *Enterprise* heroes, at least occasionally, to wait a long time to get somewhere.

Quantum fantasies can be fun, but the ultimate purpose of this new science, and of this book, is serious. It is to help us deal more effectively with our everyday reality.

WAVE-PARTICLE DUALITY AND QUANTUM MEASUREMENT

The preceding background information helps explain a couple of puzzling questions. Does the quantum picture of the electron moving in waves around the atomic nucleus imply that the electron's charge and mass are smeared all through the atom? Or does the fact that a free electron spreads out, as a wave must according to the theory of Schrödinger, mean that the electron is everywhere, with its charge now smeared all through space? In other words, how do we reconcile the wave picture of the electron with the fact that it has particlelike, localized properties? The answers are subtle.

It may seem that, with wave packets at least, we should be able to confine the electron in a small place. Alas, things do not stay that simple. A wave packet that satisfies the Schrödinger equation at a given moment in time must spread with the passage of time.

At some initial time, we may localize an electron to a tiny dot, but the electron's wave packet will spread all over town in a matter of seconds. Although initially the probability of finding the electron localized as a tiny dot is overwhelmingly high, it takes only seconds before the probability becomes considerable that the electron might appear anywhere in town. And if we wait long enough, the electron may show up anywhere in the entire country, even in the entire galaxy.

It is this spreading of the wave packet that promotes incessant jokes about quantum weirdness among the connoisseurs. For example, the quantum mechanical way of materializing a Thanksgiving turkey is as follows: Prepare your oven and wait; there is a nonzero probability that the turkey from a nearby grocery store will materialize in your oven.

Unfortunately for the turkey lover, with such massive objects as turkeys, the spreading is ever so slow. You might wait the entire lifetime of the universe to materialize even a little morsel of Thanksgiving turkey in this way.

What about the electron? How do we reconcile the spreading of the electron's wave packet all over town with the picture of a localized particle? The answer is that we must include the act of observation in our reckoning.

If we want to measure the electron's charge, we must intercept it with something like a cloud of vapor, as in a cloud chamber. As a result of this measurement, we must assume that the electron's wave collapses, so now we are able to see the electron's track through the cloud of vapor (fig. 10). According to Heisenberg: "The path of the electron comes into existence only when we observe it." When we measure it, we always find the electron localized as a particle. We may say that our measurement reduces the electron wave to the particle state.

When Schrödinger introduced his wave equation, he and others thought that perhaps they had purged physics of quantum jumps—of discontinuity—since wave motion is continuous. The particle nature of quantum objects, however, had to be reconciled with their wave nature. Thus, wave packets were introduced. Finally, with the recognition of the spreading of the wave packet and

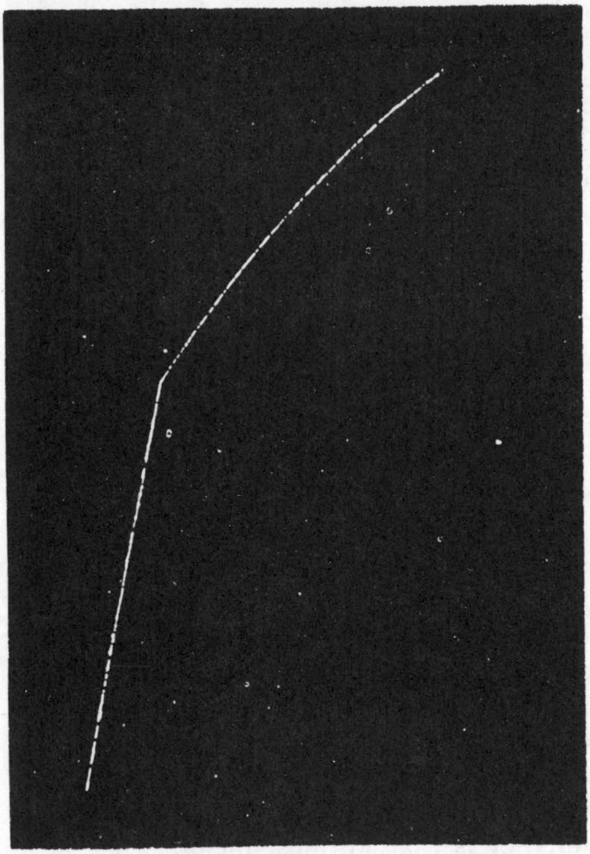

Figure 10. An electron's track through a cloud of vapor.

with the realization that it is our observation that must instantly collapse the size of the packet, we see that the collapse has to be discontinuous (continuous collapse would take time).

It seems as though we cannot have quantum mechanics without quantum jumps. Schrödinger once visited Bohr in Copenhagen, where he protested for days against quantum jumps. Ultimately, he purportedly conceded the point with this emotional outburst: "If I had known that one has to accept this damned quantum jump, I'd never have gotten involved with quantum mechanics."

Coming back to the atom, if we measure the position of the electron while it is in an atomic stationary state, we will again collapse its probability cloud to find it in a particular position, not smeared everywhere. If we make a large number of measurements to look for the electron, we will find it more often at those places where the probability of finding it is high as predicted by the

Schrödinger equation. Indeed, after a large number of measurements, if we plot the distribution of the measured positions, it will look quite like the fuzzy orbit distribution given by the solution of the Schrödinger equation (fig. 11).

How does an electron in flight appear from this perspective? When we make an initial observation of any submicroscopic projectile, we find it localized in a tiny wave packet, as a particle. After the observation, however, the packet spreads, and the spread of the packet is the cloud of our uncertainty about the packet. If we

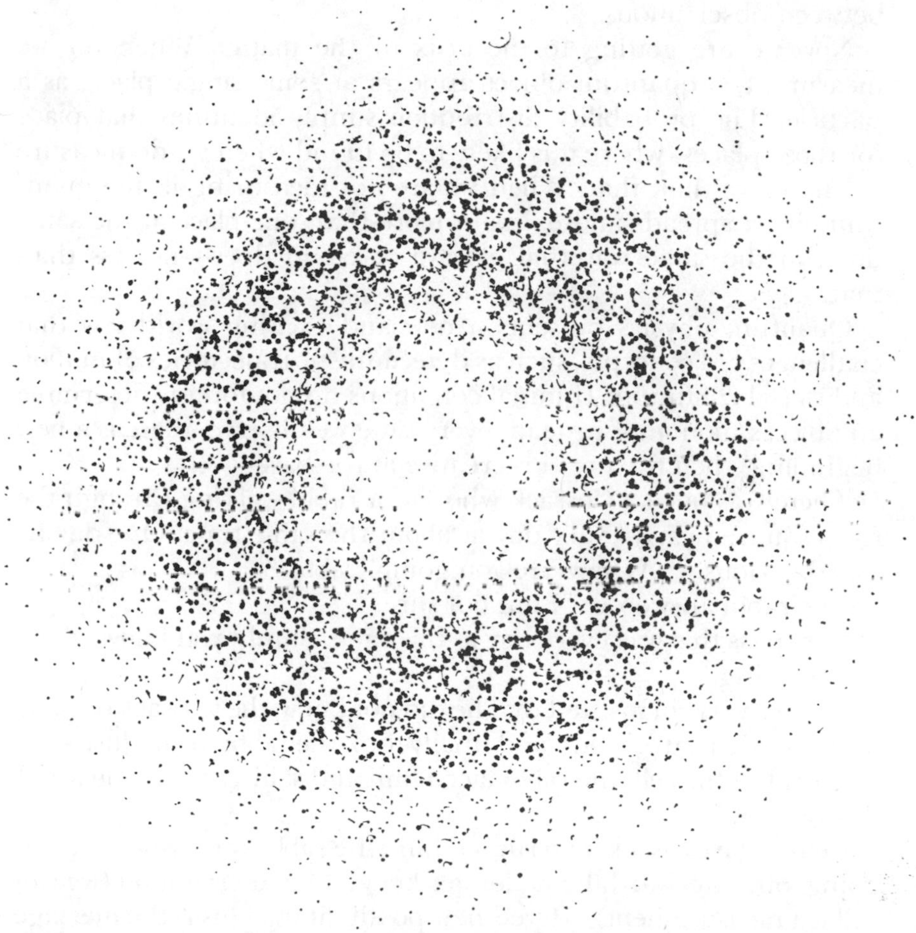

Figure 11. The results of repeated measurements of the position of a hydrogen electron in the lowest orbit. Obviously, the electron's wave usually collapses where the probability for finding it is predicted to be high, giving the fuzzy orbit.

observe again, the packet localizes once more but always spreads between our observations.

Watching electrons, said the physicist-philosopher Henry Margenau, is like watching fireflies on a summer evening. You can see a flash here and another twinkle of light there, but you have no idea where the firefly is between your observations. You cannot define a trajectory for it with any confidence. Even for a macroscopic object, such as the moon, quantum mechanics predicts essentially the same picture—the only difference being that the spreading of the wave packet is imperceptibly small (but nonzero) between observations.

Now we are getting to the crux of the matter. Whenever we measure it, a quantum object appears at some single place, as a particle. The probability distribution simply identifies that place (or those places) where it is likely to be found when we do measure it—no more than that. When we are not measuring it, the quantum object spreads and exists in more than one place at the same time, in the same way that a wave or cloud does—no less than that.

Quantum physics presents a new and exciting worldview that challenges old concepts, such as deterministic trajectories of motion and causal continuity. If initial conditions do not forever determine an object's motion, if instead, every time we observe, there is a new beginning, then the world is creative at the base level.

There was once a Cossack who saw a rabbi walking through the town square nearly every day at about the same time. One day he asked curiously: "Where are you going, rabbi?"

The rabbi answered: "I am not sure."

"You pass this way every day at this time. Surely, you know where you're going."

When the rabbi insisted that he did not know, the Cossack became irritated, then suspicious, and finally took the rabbi to jail. Just as he was locking the cell, the rabbi faced him and said gently: "You see, I didn't know."

Before the Cossack interrupted him, the rabbi knew where he was going, but afterward, he no longer knew. The interruption (we can call it a measurement) offered new possibilities. This is the message of quantum mechanics. The world is not determined by initial conditions, once and for all. Every event of measurement is potentially creative and may open new possibilities.

The Complementarity Principle

A novel way of looking at the paradox of wave-particle duality was described by Bohr. The wave and the particle natures of the electron are not dualistic, not simply opposite polarities, said Bohr. They are complementary properties revealed to us in complementary experiments. When we take a diffraction picture of an electron, we are revealing its wave nature; when we are tracking it in a cloud chamber, we are seeing its particle nature. Electrons are neither waves nor particles. We might call them "wavicles," for their true nature transcends both descriptions. This is the complementarity principle.

Since contemplating the fact that the same quantum object has such seemingly contradictory attributes as waveness and particleness can be hazardous to one's mental health, nature has provided a buffer. Bohr's complementarity principle assures us that although quantum objects have both particle and wave attributes, we can measure only one aspect of the wavicle with any given experimental arrangement at any given time. By the same token, we choose the particular aspect of the wavicle we want to see by choosing the appropriate experimental arrangement.

The Correspondence Principle

Once one has grasped the revolutionary ideas of the new physics, it would be grossly inaccurate to think that Newtonian physics is all wrong. The old physics lives on in the realm of most (but not all) bulk matter as a special case of the new physics. An important characteristic of science is that when a new order replaces an older one, it usually extends the arena to which the order applies. In the old arena, the mathematical equations of the old science still hold (having been verified by experimental data). Thus, in the domain of classical physics, the deductions of quantum mechanics for the motion of objects correspond clearly to those that are made using Newtonian mathematics, as if the bodies we were dealing with were classical. This is called the correspondence principle and was formulated by Bohr.

The relationship between classical and quantum physics in some sense resembles the optical illusion "My wife and my mother-in-law" (fig. 12). What do you see in this drawing? Initially, you see either the wife or the mother-in-law. I always see the wife first. It may actually take you a while to discover the other image in the drawing. Suddenly, if you keep at it, the other image emerges. The wife's chin transforms into the nose of the mother-in-law, her neckline into the chin of the older woman, and so on. What is going on? you may wonder. The lines are the same, but suddenly a new way of perceiving the picture has become possible for you. Very soon you find that you can easily go back and forth between the two pictures, the old and the new. You still see only one of the two images at any given time, but your consciousness has enlarged so that you are aware of the duality. In such an extended awareness, the strangeness of quantum physics begins to make sense. It even becomes exciting. Paraphrasing Hamlet's comment to Horatio, there are more things in heaven and Earth than were dreamt of in classical physics.

Figure 12. My Wife and My Mother-in-Law. (After W. E. Hill.)

Quantum mechanics gives us a wider perspective, a new context that extends our perception into a new domain. We can see nature as separate forms—either waves or particles—or we can discover complementarity: the idea that waves and particles both are inherent in the same thing.

The Copenhagen Interpretation

According to the so-called Copenhagen interpretation of quantum mechanics, developed by Born, Heisenberg, and Bohr, we calculate quantum objects as waves and interpret the waves probabilistically. We determine their attributes, such as position and momentum, somewhat uncertainly and understand them complementarily. In addition, discontinuity and quantum jumps—for example, the collapse of a sprawling wave packet upon observation—are regarded as fundamental aspects of the behavior of a quantum object. Another aspect of quantum mechanics is inseparability. Talking about a quantum object without talking about how we observe it is ambiguous because the two are inseparable. Finally, for massive macro objects, quantum mechanical predictions match those of classical physics. This introduces a suppression of such quantum effects as probability and discontinuity in the macro domain of nature that we perceive directly with our senses. Classical correspondence camouflages the quantum reality.

Cutting Through Material Realism

The principles of quantum theory make it possible to discard the unwarranted assumptions of material realism.

Assumption 1: Strong objectivity. A basic assumption that the materialist makes is that there is an objective material universe out there, one that is independent of us. This assumption has some obvious operational validity, and it is often assumed to be necessary to conduct science meaningfully. Is this assumption really valid? The lesson of quantum physics is that we choose which aspect—wave or particle—a quantum object is going to reveal in a given situation. Moreover, our observation collapses the quantum wave packet to a localized particle. Subjects and objects are inextricably blended together. If subjects and objects mesh in this way, how can we uphold the assumption of strong objectivity?

Assumption 2: Causal determinism. Another assumption of the classical scientist that lends credence to material realism is that the world is fundamentally deterministic—all we have to know are the forces acting on each object and the initial conditions (the initial velocity and position of the object). The quantum uncertainty principle, however, says that we can never determine both an object's velocity and its position simultaneously with absolute accuracy. There will always be error in our knowledge of the initial conditions, and strict determinism does not prevail. The idea of causality itself is even suspect. Since the behavior of quantum objects is probabilistic, a strict cause-effect description of the behavior of a single object is impossible. Instead, we have statistical cause and statistical effect when talking about a large group of particles.

Assumption 3: Locality. The assumption of locality—that all interactions between material objects are mediated via local signals—is crucial to the materialistic view that objects exist essentially independent and separate from one another. If, however, waves spread over vast distances and then instantly collapse when we take measurements, then the influence of our measurement is not traveling locally. Thus locality is ruled out. This is another fatal blow to material realism.

Assumptions 4 and 5: Materialism and epiphenomenalism. The materialist maintains that subjective mental phenomena are but epiphenomena of matter. They can be reduced to material brain stuff alone. In order to understand the behavior of quantum objects, however, we seem to need to inject consciousness—our ability to choose—according to the complementarity principle and the idea of subject-object mixing. Moreover, it seems absurd that an epiphenomenon of matter can affect matter: If consciousness is an epiphenomenon, how can it collapse the spread-out wave of a quantum object to a localized particle when it takes a quantum measurement?

The correspondence principle notwithstanding, the new paradigm of physics—quantum physics—contradicts the dicta of material realism. There is no way around this conclusion. We cannot say, citing correspondence, that classical physics holds for macro objects for all practical purposes and that since we live in the macro world, we will assume that the quantum strangeness confines itself to the submicroscopic domain of nature. On the contrary, the strangeness haunts us all the way to the macro level. There are unresolvable quantum paradoxes if we divide the world into domains of classical and quantum physics.

In India, people ingeniously catch monkeys with a jar of chickpeas. The monkey reaches into the jar to grab a fistful of chickpeas. Alas, with its fist closed on the food, it can no longer remove its hand. The mouth of the jar is too small for its fist. The trap works because the monkey's greed prohibits him from letting go of the chickpeas. The axioms of material realism—materialism, determinism, locality, and so forth—served us well in the past when our knowledge was more limited than it is today, but now they have become our trap. We may have to let go of the chickpeas of certainty in order to embrace the freedom that lies outside the material arena.

If material realism is not an adequate philosophy for physics, what philosophy can deal with all the strangeness of quantum behavior? It is the philosophy of monistic idealism, which has been the basis of all religions worldwide.

Traditionally, only religions and the humanistic disciplines have given value to human life beyond physical survival—value through our love of aesthetics; our creativity in art, music, and thought; and our spirituality in the intuition of unity. The sciences, locked into classical physics and its philosophical baggage of material realism, have been the Pied Piper of skepticism. Now the new physics is crying out for a new, liberating philosophy—one befitting our current level of knowledge. If monistic idealism fits the need, for the first time since Descartes, science, the humanities, and the religions can walk arm-in-arm in the search for the whole human truth.

Chapter 4

THE PHILOSOPHY OF MONISTIC IDEALISM

THE ANTITHESIS OF MATERIAL REALISM is monistic idealism. In this philosophy, consciousness, not matter, is fundamental. Both the world of matter and the world of mental phenomena, such as thought, are determined by consciousness. In addition to the material and the mental spheres (which together form the immanent reality, or world of manifestation), idealism posits a transcendent, archetypal realm of ideas as the source of material and mental phenomena. It is important to recognize that monistic idealism is, as its name implies, a unitary philosophy; any subdivisions, such as the immanent and the transcendent, are within consciousness. Thus consciousness is the only ultimate reality.

In the West, the philosophy of monistic idealism has been stated most influentially by Plato, who in *The Republic* gave us his famous allegory of the cave.[1] As hundreds of generations of philosophy students have learned, this allegory clearly illustrates the fundamental concepts of idealism. Plato imagines human beings sitting in a cave in a fixed position so that they always face the wall. The great universe outside is a shadow show projected on the wall of the cave, and we humans are shadow watchers. We watch shadow-illusions that we mistake for reality. The real reality is behind us, in the light and archetypal forms that cast the shadows on the wall. In this allegory, the shadow shows are the unreal immanent manifestations in human experience of archetypal realities that belong to a transcendent world. In truth, light is the only reality, for light is all we see. In monistic idealism, consciousness is like the light in Plato's cave.

The same essential ideas occur repeatedly in the idealist literature of many cultures. In the Vedanta literature of India, the Sanskrit word *nama* is used to denote transcendent archetypes, and *rupa* signifies their immanent form. Beyond nama and rupa shines the light of the *Brahman*, the universal consciousness, the one without a second, the ground of all being. "This entire universe of which we speak and think is nothing but Brahman. Brahman dwells beyond the range of Maya [illusion]. There is nothing else."[2]

In Buddhist philosophy, the material and idea realms are referred to as *Nirmanakaya* and *Sambhogakaya* respectively, but beyond these is the light of one consciousness, *Dharmakaya*, which illuminates both. And in reality, there is only Dharmakaya. "Nirmanakaya [is] the appearance body of Buddha and his inscrutable activities. Sambhogakaya possesses vast and boundless potentiality. Buddha's Dharmakaya is free from any perception or conception of form."

Perhaps the Taoist symbol of yin and yang (fig. 13) is more generally familiar than are the Indian symbols. The light yang, regarded as a male symbol, defines the transcendent realm, and the dark yin, regarded as a female symbol, defines the immanent.

Figure 13. The yin-yang symbol.

Note their figure-ground relationship. "That which lets now the dark, now the light appear is the *Tao*," the one that transcends its complementary manifestations.

Similarly, the Jewish Kabbalah describes two orders of reality: the transcendent one, represented by the Sefiroth as Theogony; and the immanent one, which is the *alma de-peruda*, the "world of separation." According to the *Zohar*, "if one contemplates things in mystical meditation, everything is revealed as one."

In the Christian world, the names of the transcendent and immanent realms—heaven and earth—are part of our everyday vocabulary. However, our everyday usage fails to recognize the origin of these notions in monistic idealism. Beyond the kingdoms of heaven and earth there is the Godhead, the King of the kingdoms. The kingdoms do not exist separately from the King: The King is the kingdoms. Writes the Christian idealist Dionysius: "It [consciousness—the ground of being] is within our intellects, souls and bodies, in heaven, on earth, and whilst remaining the same in Itself. It is at once in, around and above the world, super-celestial, superessential, a sun, a star, fire, water, spirit, dew, cloud, stone, rock, all that is."[3]

In all these descriptions, note that the one consciousness is said to come to us through complementary manifestations: ideas and forms, nama and rupa, Sambhogakaya and Nirmanakaya, yang and yin, heaven and earth. This complementary description is an important facet of idealist philosophy.

When we look around us, ordinarily we see only matter. Heaven is not a tangible object of ordinary perception. This is not only what leads us to refer to matter as real but also what induces us to accept a realist philosophy that proclaims matter (and its alternate form, energy) to be the only reality. Many idealists have maintained, however, that it is possible to experience heaven directly if one seeks beyond mundane everyday experiences. People who make such claims are known as mystics. Mysticism offers experiential proof of monistic idealism.

MYSTICISM

Realism grew out of our everyday perceptions. In our everyday experiences of the world, evidence abounds that things are material and separate from each other and from us.

Of course, mental experiences do not fit neatly into such a formulation. Mental experiences, such as thought, do not seem to be material, so we have developed a dualistic philosophy that relegates mind and body to separate domains. The shortcomings of dualism are well known. Notably, it cannot explain how a separate, nonmaterial mind interacts with a material body. If there are such mind-body interactions, then there have to be exchanges of energy between the two domains. In myriad experiments, we find that the energy of the material universe by itself remains a constant (this is the law of conservation of energy). Neither has any evidence shown that energy is lost to or gained from the mental domain. How can that be if there are interactions going on between the two domains?

Idealists, although they hold consciousness to be the primary reality and thus give value to our subjective, mental experiences, do not propose that consciousness is mind. (Beware of possible semantic confusion: *Consciousness* is a relatively recent word in the English language. The word *mind* is often used to denote consciousness, especially in the older literature. In this book, the distinction between the concepts of mind and consciousness is necessary and important.) Instead, they propose that material objects (such as a ball) and mental objects (such as the thought of a ball) are both objects in consciousness. In an experience there is also the subject, the experiencer. What is the nature of this experiencer? This is a question of utmost importance in monistic idealism.

According to monistic idealism, the consciousness of the subject in a subject-object experience is the same consciousness that is the ground of all being. Therefore, consciousness is unitive. There is only one subject-consciousness, and we are that consciousness. "Thou art that!" say Hindu holy books known collectively as the Upanishads.

Why, then, in our ordinary experience do we feel so separate? This separateness, insists the mystic, is an illusion. If we meditate on the true nature of our self, we shall find, as mystics from many ages and times have found, that there is only one consciousness behind all the diversity. This one consciousness/subject/self goes by many names. Hindus refer to it as the atman; Christians call it the Holy Spirit, or in Quaker Christianity, the inner light. By whatever name it is called, all agree that the experience of this one consciousness is of inestimable value.

Buddhist mystics often refer to the consciousness beyond the individual as the no-self, which leads to the potential confusion that they may be negating consciousness altogether. But the Buddha clarified this misconception: "There is an Unborn, Unoriginated, Uncreated, Unformed. If there were not this Unborn, this Unoriginated, this Uncreated, this Unformed, escape from the world of the born, the originated, the created, the formed would not be possible. But since there is an Unborn, Unoriginated, Uncreated, Unformed, therefore is escape possible from the world of the born, the originated, the created, the formed."[4]

Mystics, then, are those people who offer testimony to this fundamental reality of unity in diversity. A sampling of mystical writings from different cultures and spiritual traditions bears witness to the universality of the mystical experience of unity.[5]

The Christian mystic Catherine Adorna of Genoa, who lived in fifteenth-century Italy, simply and beautifully stated her knowledge: "My being is God, not by simple participation, but by a true transformation of my being."[6]

The great Hui-Neng of sixth-century China, an illiterate peasant whose sudden illumination eventually resulted in the founding of Zen Buddhism, declared: "Our very self-nature is Buddha, and apart from this nature there is no other Buddha."[7]

The twelfth-century Sufi mystic Ibn al-Arabi, revered by Sufis as the Sheikh of sheikhs, had this to say: "Thou art neither ceasing to be nor still existing. Thou art He, without one of these limitations. Then if thou know thine own existence thus, then thou knowest God; and if not, then not."[8]

The fourteenth-century Kabbalist Moses de Leon, probably the author of the *Zohar*, which is the primary sourcebook for Kabbalists, wrote: "God . . . when he has just decided to launch upon his work of creation is called *He*. God in the complete unfolding of his Being, Bliss and Love, in which he becomes capable of being perceived by the reasons of the heart . . . is called *You*. But God, in his supreme manifestation, where the fullness of His Being finds its final expression in the last and all-embracing of his attributes, is called *I*."[9]

The eighth-century mystic Padmasambhava is credited with bringing Tantric Buddhism to Tibet. His consort, the charismatic Yeshe Tsogyel, expressed her wisdom this way: "But when you finally discover me, the one naked Truth arisen from within, Absolute Awareness permeates the Universe."[10]

Meister Ekhart, the thirteenth-century Dominican monk, wrote: "In this breaking-through I receive that God and I are one. Then I am what I was, and then I neither diminish nor increase, for I am then an immovable cause that moves all things."[11]

From the tenth-century Sufi mystic Monsoor al-Halaj came the pronouncement: "I am the Truth!"[12]

The eighth-century Hindu mystic Shankara exuberantly expressed his realization: "I am reality without beginning, without equal. I have no part in the illusion of 'I' and 'you,' 'this' and 'that.' I am Brahman, one without a second, bliss without end, the eternal, unchanging truth. . . . I dwell within all beings as the soul, the pure consciousness, the ground of all phenomena, internal and external. I am both the enjoyer and that which is enjoyed. In the days of my ignorance, I used to think of these as being separate from myself. Now I know that I am All."[13]

And finally, Jesus of Nazareth declared: "My father and I are one."[14]

What is the value of the experience of unity? For the mystic, it opens the door to a transformation of being that liberates love, universal compassion, and freedom from the bondage of living in acquired separateness and from the compensating attachments to which we cling. (This liberated being is called *moksha* in Sanskrit.)

The idealist philosophy grew out of the experiences and creative intuitions of mystics, who constantly stress the direct experiential aspect of the underlying reality. "The Tao that can be spoken is not the absolute Tao," said Lao Tzu. The mystics caution that all teachings and metaphysical writings must be regarded as fingers pointing to the moon rather than as the moon itself.

As the *Lankavatara Sutra* reminds us: "These teachings are only a finger pointing to the Noble wisdom. . . . They are intended for the consideration and guidance of the discriminating minds of all people, but they are not the Truth itself, which can only be self-realized within one's own deepest consciousness."[15]

Alternatively, some mystics resort to paradoxical descriptions. Writes Ibn al-Arabi: "It [consciousness] is neither attributed with being nor with nonbeing. . . . It is neither existent nor non-existent. It is not said to be either the First or the Last."[16]

Indeed, the idealist metaphysics itself can be seen to be paradoxical, involving, as it does, the paradoxical concept of transcendence. What is transcendence? The philosophy can only say, neti, neti—not this, not that. But what is it? The philosophy remains silent. Or, alternatively, says one of the Upanishads: "It is within all this/It is outside all this."[17]

Is the transcendent realm within the immanent world? Yes. Is it outside the immanent world? Yes. It gets very confusing.

The idealist philosophy also remains largely silent in answering such questions as, How does the undivided consciousness divide itself into subject-object reality? How does the one consciousness become many? Saying that the observed multiplicity of the world is illusion hardly satisfies us.

In this book, we will argue that monistic idealism is the correct philosophy for science in view of quantum physics. The integration of science and mysticism also helps resolve some of the difficult questions raised by mysticism.

The integration of science and mysticism should not be too disconcerting. After all, they share an important similarity: Both grew out of empirical data interpreted in the light of theoretical explanatory principles. In science, theory serves both as explanation of data and as the instrument of prediction and guidance for future experiments. The idealist philosophy, too, can be viewed as a creative theory that acts as an explanation of empirical observations of the mystics as well as guidance for other researchers of Truth. Finally, like science, mysticism seems to be a universal enterprise. There is no parochialism in mysticism. Parochialism enters when religions simplify mystical teachings to make them more communicable to the masses of humankind.

RELIGION

To arrive at an understanding of Truth, a mystic usually discovers and employs a particular methodology. The methodologies, or spiritual paths, have both similarities and differences. The differences, which are secondary to the universality of the mystical insight itself, contribute to the differences in the religions founded on the teachings of the mystics. For example, Buddhism developed from the teachings of the Buddha, Judaism from the teachings of Moses, Christianity from those of Jesus, Islam from those of Mohammed (although strictly speaking, Mohammed is regarded as the last of a whole lineage of prophets, including Moses and Jesus), and Taoism from the teachings of Lao Tzu. This rule, however, is not without exception. Hinduism is not based on the teachings of a particular teacher but instead encompasses many paths, many teachings.

Mysticism involves a search for the truth about ultimate reality, but the function of the religion is somewhat different. The followers of a particular mystic (most often after the mystic's death) may recognize that the individual search for truth is not for everyone. Most people, lost in the illusion of their ego-separateness and busy in its pursuits, are not motivated to discover the truth themselves. How, then, can the light of the mystic's realization be shared with these people?

The answer is, by simplifying it. The followers simplify the truth to make it accessible to the average person. Such a person is usually caught up in the demands of daily life. Lacking the time and devotion necessary to understand the subtlety of transcendence, he or she cannot appreciate the importance of direct mystical experience. So, the purveyors of the mystic's truth replace direct experience of unitive consciousness with the idea of God. Unfortunately, God, the transcendent creator of the immanent world, is recast in the ordinary person's mind into the dualistic image of a mighty King in Heaven who rules the Earth below. Unavoidably, the mystic's message is diluted and distorted.

The mystic's well-meaning followers inadvertently play the role of the devil in an old joke: God and the devil were walking together when God picked up a piece of paper. "What does it say?" the devil inquired. "Truth," said God serenely. "Give it to me," said the devil eagerly. "I'll organize it for you."

Yet, despite the difficulties and fallibilities of organization, the religion does convey the spirit of the mystic's message; this is what gives religion its vitality. After all, the value to mystics of realizing the transcendent nature of Reality is that they become secure in a mode of being in which such virtues as love become simple. How can you not love when there is one consciousness and you know that you and the other are not really separate?

But how does one motivate an ordinary person who does not realize oneness to love others? The mystic clearly sees that ignorance of the transcendent oneness is the barrier to love. The net effect of the absence of love is suffering. To avoid suffering, counsels the mystic, we must turn inward and commence the journey to self-realization. In the religious context, this teaching is translated into the dictum that if we are to redeem ourselves, we must turn to God as the supreme value in our lives. The method of this redemption is a set of practices, based on the original teachings, that forms the moral code of the particular religion—the ten commandments and the Golden Rule of Christian ethics, the Buddhist precepts, the Koranic or Talmudic law, and so forth.

Of course, not all religions introduce the concept of God. In Buddhism, for example, there is no concept of God. On the other hand, in Hinduism there are many gods. Even in these cases, however, the above considerations of religion are evident. Thus we arrive at three universal aspects of all exoteric religions:

1. All religions start with the premise that there is a wrongness in the way we are. The wrongness is variously called ignorance, original sin, evil, or just suffering.
2. All religions promise an escape from this wrongness, provided the "way" is followed. The escape is variously called salvation, liberation from the wheel of suffering in the world, enlightenment, or an eternal life in the kingdom of God, heaven.
3. The way consists of taking refuge in the religion and the community formed by the followers of the religion and following a prescribed code of ethics and social rules. Aside from how the esoteric teaching of transcendence is compromised, it is in the codes of ethics and social rules that the various religions differ from one another.[18]

Notice the essential dualism in the first premise: wrong and right (or evil and good). In contrast, the mystical journey consists in transcending all dualities, including the one of evil and good. Also notice that the second premise is turned by the clergy into carrots and sticks—heaven and hell. Mysticism, on the other hand, does not dichotomize heaven and hell; both are natural concomitants of how we live.

As you can see, when filtered by the world's religions, the monism of monistic idealism becomes ever more obscure, and dualistic ideas prevail. In the East, thanks to an endless supply of students of mysticism, monistic idealism in its esoteric form has popularly retained at least some passing familiarity and respect. In the West, however, mysticism has had relatively little impact. The dualism of the Judeo-Christian monotheistic religions has dominated the popular psyche, supported by a powerful hierarchy of interpreters. Like mind-body Cartesian dualism, however, the dualism of God and the world does not seem to hold up to scientific scrutiny.[19] As scientific data undermine religion, there is a tendency to throw out the baby with the bathwater—the baby being the ethics and values that the religion teaches, ethics and values that continue to have validity and usefulness.

Exposing the illogic of dualistic religions need not result in the monistic philosophy of material realism. As we have seen, an alternative monism is available. In view of the way that quantum physics has demolished material realism, monistic idealism may be the only viable monistic philosophy of reality. The other option is to give up on metaphysics entirely, which for a while was the direction in philosophy. The trend now seems to be reversing.

Now we must face the crucial question: Is science compatible with monistic idealism? If not, we must abandon metaphysics when doing science, adding to the looming crisis of faith. If yes, we must reformulate science in accordance with the demands of philosophy. In this book, we argue that monistic idealism is not only compatible with quantum physics but even essential to its interpretation. The paradoxes of the new physics disappear when we examine them from the point of view of monistic idealism. Furthermore, quantum physics combined with monistic idealism gives us a powerful paradigm with which we can resolve some of the paradoxes of mysticism, such as the question of transcendence and plurality. Our work points toward the beginnings of an idealist science and the revitalization of religions.

Idealist Metaphysics for Quantum Objects

Quantum objects show the complementary aspects of wave and particle. Is quantum complementarity—the solution of the wave-particle duality—the same as the complementarity of monistic idealism?

The writer George Leonard obviously saw a parallel between the two types of complementarity when he wrote in *The Silent Pulse*: "Quantum mechanics is the ultimate koan of our times." Koans are the Zen Buddhists' tool for breaking through apparent paradoxes to transcendent solutions. Let us compare koans with complementarity.

In one koan, Zen aspirant Daibai asked Baso, the Zen master, "What is Buddha?" Baso answered, "This mind is Buddha."

Another monk asked the same question, "What is Buddha?" Baso replied, "This mind is not Buddha."

Now compare this with Bohr's complementarity. Ask Bohr, "Is the electron a particle?" Sometimes Bohr will reply, "It is." When you look at the cloud chamber track of an electron, it makes sense to say that an electron is a particle. Looking at the diffraction pattern of electrons, however, Bohr will say, mischievously smoking his pipe: "You must agree that an electron is a wave." It seems that, like the Zen master Baso, Bohr is of two minds regarding the nature of electrons.

Quantum waves are waves of probability. We need to experiment with many wavicles to see the wave aspect, as in the diffraction pattern. *We never, never see the wave aspect of a single quantum object; experimentally, a single wavicle always, always reveals itself as a localized particle.* The wave aspect nevertheless persists even for a single wavicle. Does the wave aspect of a single wavicle exist in a transcendental space, since it never manifests in ordinary space? Is Bohr's idea of complementarity pointing to the same transcendent order of reality that the philosophy of monistic idealism proposes?

Bohr never said yes in so many words to such questions, and yet his coat of arms displays the Chinese symbol of yin and yang. (He was knighted in 1947.) Can it be that Bohr understood the complementarity of quantum physics in a way similar to monistic idealism, that he supported an idealist metaphysics for quantum objects?

Recall the uncertainty principle. If the product of the uncertainty in position and the uncertainty in momentum is a constant, then reducing the uncertainty of one measure increases the uncertainty of the other measure. Extrapolating from this argument, we can see that if the position is known with complete certainty, then the momentum becomes completely uncertain. And vice versa. When the momentum is known with complete certainty, the position becomes completely uncertain.

Many quantum initiates protest these implications of the uncertainty principle. "But surely," they say, "the electron must be somewhere; we just don't know where." No, it is worse. We cannot even define the position of the electron in ordinary space and time. Obviously, quantum objects exist very differently from the familiar macro objects of everyday life.

Heisenberg also recognized that a quantum object cannot occupy a given place and still be moving at the same time in a predictable fashion. Any attempt to take a snapshot of a submicroscopic object results only in giving us its position, but we lose information about its state of motion. And vice versa.

This observation raises another question. What is the object doing between snapshots? (This is similar to the question about electrons making quantum jumps between Bohr orbits: Where does the electron go between jumps?) We cannot assign an electron a trajectory. To do that we would need to know both the electron's position and its velocity at some initial time, and that would violate the uncertainty principle. Can we assign the electron any manifest reality in space and time between observations? According to the Copenhagen interpretation of quantum mechanics, the answer is no.

Between observations, the electron spreads out in accordance with the Schrödinger equation, but probabilistically, in potentia, said Heisenberg, who adopted the word *potentia* from Aristotle.[20] Where do these potentia exist? Since the electron's wave collapses immediately upon our observation, potentia could not be within the material domain of space-time; in space-time all objects have to obey the Einsteinian speed limit, remember? Thus the domain of potentia must be outside space-time. Potentia exist in a transcendent domain of reality. Between observations, the electron exists as a possibility form, like a Platonic archetype, in the transcendent domain of potentia. ("I dwell in Possibility," wrote the poet Emily Dickinson. If the electron could talk, this is how it would likely describe itself.)

Electrons are too remote from ordinary personal reality. Suppose we ask, Is the moon there when we are not looking at it? To the extent that the moon is ultimately a quantum object (being composed entirely of quantum objects), we must say no—so says physicist David Mermin.[21] Between observations, the moon also exists as a possibility form in transcendent potentia.

Perhaps the most important, and the most insidious, assumption that we absorb in our childhoods is that of the material world of objects existing out there—independent of subjects, who are the observers. There is circumstantial evidence in favor of such an assumption. Whenever we look at the moon, for example, we find the moon where we expect it along its classically calculated trajectory. Naturally we project that the moon is always there in space-time, even when we are not looking. Quantum physics says no. When we are not looking, the moon's possibility wave spreads, albeit by a minuscule amount. When we look, the wave collapses instantly; thus the wave could not be in space-time. It makes more sense to adapt an idealist metaphysic assumption: There is no object in space-time without a conscious subject looking at it.

So quantum waves are like Platonic archetypes in the transcendent domain of consciousness, and the particles that manifest upon our observation are the immanent shadows on the cave wall. Consciousness is the agency that collapses the wave of a quantum object, which exists in potentia, making it an immanent particle in the world of manifestation. This is the basic idealist metaphysics that we shall use for quantum objects in this book. Under the illumination of this simple idea, we shall see all the famous paradoxes of quantum physics vanish like morning mist.

Note that Heisenberg himself almost came up with the idealist metaphysic when he introduced the concept of potentia. The important new element is that the domain of potentia also exists in consciousness. Nothing is outside consciousness. This monistic view of the world is crucial.

SCIENCE DISCOVERS TRANSCENDENCE

Until the present interpretation of the new physics, the word *transcendence* was seldom mentioned in the vocabulary of physics. The term was even considered heretical (and still is, to some extent) to the classical-law-abiding practitioners of a deterministic, cause-and-effect science in a clockwork universe.

To the Roman philosophers of antiquity, transcendence meant "the state of extending or lying beyond the limits of all possible experience and knowledge," or "being beyond comprehension." To monistic idealists, similarly, transcendence means *not this, not anything known*. Today, modern science is venturing into realms that for more than four millennia have been the fiefdoms of religion and philosophy. Is the universe only an objectively predictable series of phenomena that humankind can observe and control, or is it much more elusive and even more wonderful? During the past three hundred years, science has become the unrivaled benchmark of reality. We are privileged to be a part of this evolutionary and transcendent process by which science is changing not only itself but also our perspective on reality.

A tantalizing development—an experiment by a team of physicists at Orsay, France[22]—not only has confirmed the idea of transcendence in quantum physics but is also clarifying the concept of transcendence. The experiment by Alain Aspect and his collaborators directly shows that when two quantum objects are correlated, if we measure one (thus collapsing its wave function), the other's wave function is instantly collapsed as well—even at a macroscopic distance, even when there is no signal in space-time to mediate their connection. Einstein, however, proved that all connections and interactions in the material world must be mediated by signals traveling through space (the locality principle) and thus must be limited by the speed of light. Where, then, exists the instantaneous connection between correlated quantum objects that is responsible for their signal-less action at a distance? The succinct answer is: in the transcendent domain of reality.

The technical name for signal-less, instantaneous action at a distance is nonlocality. The correlation of quantum objects observed in Aspect's experiment is a nonlocal correlation. Once we accept quantum nonlocality as an established physical aspect of the world in which we live, it becomes easier within science to conceive of a transcendent domain outside the manifest physical domain of space-time. According to the physicist Henry Stapp, the message of quantum nonlocality is that "the fundamental process of Nature lies outside space-time but generates events that can be located in space-time."[23]

Caution: If "outside space" makes you think of another "box" outside the spatial "box" we are in, forget it. The other box can be made as much a part of the universe of space as ours, by definition. With nonlocal connection, we are forced to conceptualize a domain of reality outside space-time because a nonlocal connection cannot happen in space-time.

There is another paradoxical way to think of the nonlocal reality—as being both everywhere and nowhere, everywhen and nowhen. This is still paradoxical, but it is suggestive, isn't it? I cannot resist playing on the word *nowhere,* which as a child I read (the first time I encountered it) as *now here.* Nonlocality (and transcendence) is nowhere and now here.

Democritus, some 2,500 years ago, propounded the philosophy of materialism, but shortly thereafter Plato gave us one of the first clear statements of the philosophy of monistic idealism. As Werner Heisenberg noted, quantum mechanics indicates that, of the two minds, Plato and Democritus, that have most influenced Western civilization, Plato may be the ultimate winner.[24] The success that Democritus's materialism has enjoyed in science for the past three hundred years may be only an aberration. Quantum theory interpreted according to an idealist metaphysics is paving the road for an idealist science in which consciousness comes first and matter pales to secondary importance.

PART 2

IDEALISM AND THE RESOLUTION OF THE QUANTUM PARADOXES

Habits of thought die hard. Although quantum mechanics has replaced classical mechanics as the fundamental theory of physics, many physicists, conditioned by the old worldview, still find the idealist implications of quantum mechanics difficult to entertain. They do not want to ask the embarrassing metaphysical questions raised by quantum mechanics. They hope that if such problems are ignored, they will go away. Once, at the beginning of a discussion of paradoxes in quantum mechanics, Nobel laureate Richard Feynman caricatured this attitude in his inimitable tongue-in-cheek manner: "Hush, hush," he said. "Close the doors."

In the next five chapters we shall open the doors and unabashedly expose the paradoxes of quantum physics. Our purpose will be to show that, when viewed with the light of monistic idealism, the quantum paradoxes turn out not to be so shocking or paradoxical after all. A strict adherence to an idealist metaphysics, one based on a transcendent, unitive consciousness collapsing the quantum wave, resolves in a nonarbitrary fashion all the paradoxes of quantum physics. We shall find that it is completely possible to do science within the framework of monistic idealism. The result is an idealist science that integrates spirit and matter.

The idea that consciousness collapses the quantum wave was originally proposed by the mathematician John von Neumann in the 1930s. What took us so long to take this idea seriously? Perhaps a brief discussion of how my own clarity on this issue developed will help.

One of the difficulties I had with von Neumann's proposal had to do with experimental data. When we look, we seem to be always conscious. Then the question of consciousness collapsing the quantum waves seems purely academic. Could one ever find a situation where one is looking, but is not conscious? Notice how paradoxical this sounds.

In 1983, I was invited to a ten-week-long seminar on consciousness at the psychology department at the University of Oregon. I was particularly flattered that these erudite psychologists patiently listened to six full hours of talks that I gave on the quantum ideas. The real reward came, however, when one of the graduate students of psychologist Michael Posner's group reported some cognitive data collected by a fellow named Tony Marcel. Some of the data concerned "unconscious seeing": exactly what I was looking for.

With heart palpitating, I listened to the data and relaxed only when I realized that the data are completely in agreement with consciousness collapsing the quantum state of the brain-mind when we see consciously (see chapter 7). In unconscious seeing, there is no collapse, and that really made a lot of experimental difference. Soon I realized also how to resolve the slight paradox that the distinction of conscious and unconscious perception creates. The trick is to distinguish between consciousness and awareness.

Chapter 5

OBJECTS IN TWO PLACES AT ONCE AND EFFECTS THAT PRECEDE THEIR CAUSES

THE FUNDAMENTAL TENETS of material realism simply do not hold up. In place of causal determinism, locality, strong objectivity, and epiphenomenalism, quantum mechanics offers probability and uncertainty, wave-particle complementarity, nonlocality, and mixing of subjects and objects.

About the probability interpretation of quantum mechanics, which breeds uncertainty and complementarity, Einstein used to say that God does not play dice. To see what he meant imagine that you are doing an experiment with a radioactive sample that, of course, obeys probabilistic quantum laws of decay. Your job is to measure the time it takes for ten radioactive events—ten clicks of your Geiger counter. Suppose further that it takes on the average half an hour for the ten cases of radioactive decay to occur. Behind that average lurks probability. Some runs could take thirty-two minutes, other runs twenty-five minutes, and so on. To complicate things, you have a bus to catch to meet your fiance, who absolutely hates to be kept waiting. And guess what? Your last run takes forty minutes because a single atom, at random, will not decay like the average ones did. So you miss your bus, your fiance breaks up with you, and your life is ruined.[1] This may be a somewhat silly, concocted example of what happens in a world whose God plays dice, but it does make the point. Probabilistic events can be depended on only on the average.

The randomness of atomic events—the diciness of fate, as it were—is abhorrent to a determinist. The determinist thinks about probability in the way in which we think of it in classical physics and in everyday life: It is a characteristic of large ensembles of objects—ensembles so large and intricate that we cannot, as a practical matter, predict them, though such prediction is possible in principle. To the determinist, probability is simply a convenience of thought; the physical laws that guide the motions of individual objects are completely determined and therefore completely predictable. It was Einstein's belief that the quantum mechanical universe is also this way: There are hidden variables behind the quantum uncertainties. The probabilities of quantum mechanics are simply matters of convenience. If such were the case, quantum mechanics would have to be a theory of ensembles. Indeed, if we do not apply the probability wave description to a single quantum object, then we do not get into the paradoxes that excite us—wave-particle complementarity and the inseparability of the quantum object from considerations of its observation.

Unfortunately, things are not that simple. Considering a couple of quantum-mechanical experiments will show how hard it is to rationalize away the paradoxes of quantum physics.

THE DOUBLE-SLIT EXPERIMENT

We can never see the wave aspect of a single wavicle. Whenever we look, all we see is the localized particle. Should we, therefore, assume that the solution is transcendent metaphysics? Or should we abandon the idea that there is a wave aspect of a single wavicle? Perhaps the waves that appear in quantum physics are a characteristic only of groups or ensembles of objects.

To determine whether this is so, we can analyze an experiment commonly used to study wave phenomena: the double-slit experiment. In the setup for this experiment, a beam of electrons passes through a screen that has two narrow slits in it (fig. 14). Since electrons are waves, the beam is split into two sets of waves by the two-slitted screen. These waves then interfere with one another, and the result of the interference shows on a fluorescent screen.

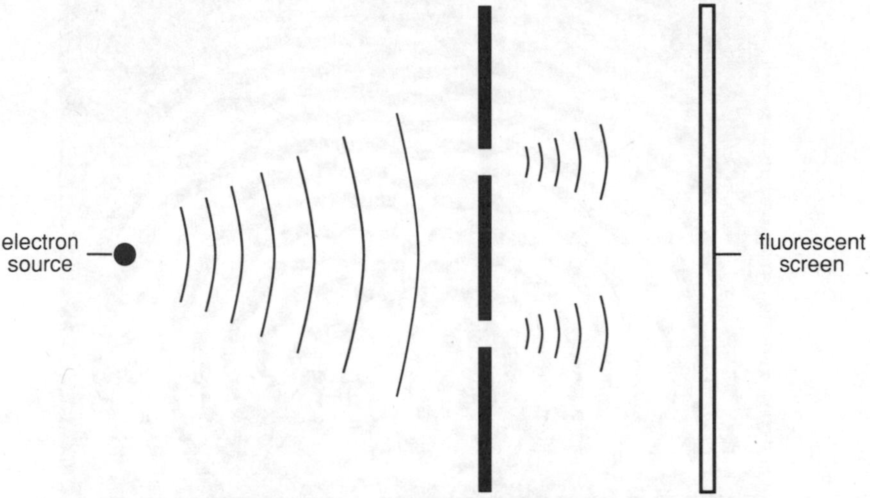

Figure 14. The double-slit experiment for electrons.

Simple enough? Let me review the phenomenon of wave interference. For an easy demonstration, if you are not familiar with the interference phenomenon, stand in a bathtub filled with water and make two water-wave trains by rhythmically marching in place. The waves will make an interference pattern (fig. 15a). At some points they will reinforce each other (fig. 15b); at other points they will cause mutual destruction (fig. 15c). Hence, the pattern.

Similarly, there are places on the fluorescent screen where the electron waves from the two slits arrive in phase, matching their dance steps. At these places their amplitudes add, and the total wave is reinforced. In between these bright spots there are places where the two waves arrive out of phase and cancel each other out. The result of this constructive and destructive interference, then, shows on the fluorescent screen as a pattern of alternating bright and dark fringes: an interference pattern (fig. 16). Importantly, the spacing of the fringes enables us to measure the wavelength of the waves.

Remember, though, that the electron waves are probability waves. Thus we must say that it is the probability of an electron arriving at the light areas that is high, and the probability of an electron arriving at the dark areas that is low. We must not get carried away and conclude from the interference pattern that the electron waves are classical waves, because the electrons do arrive at the fluorescent screen in a very particlelike way: one localized flash per electron. It

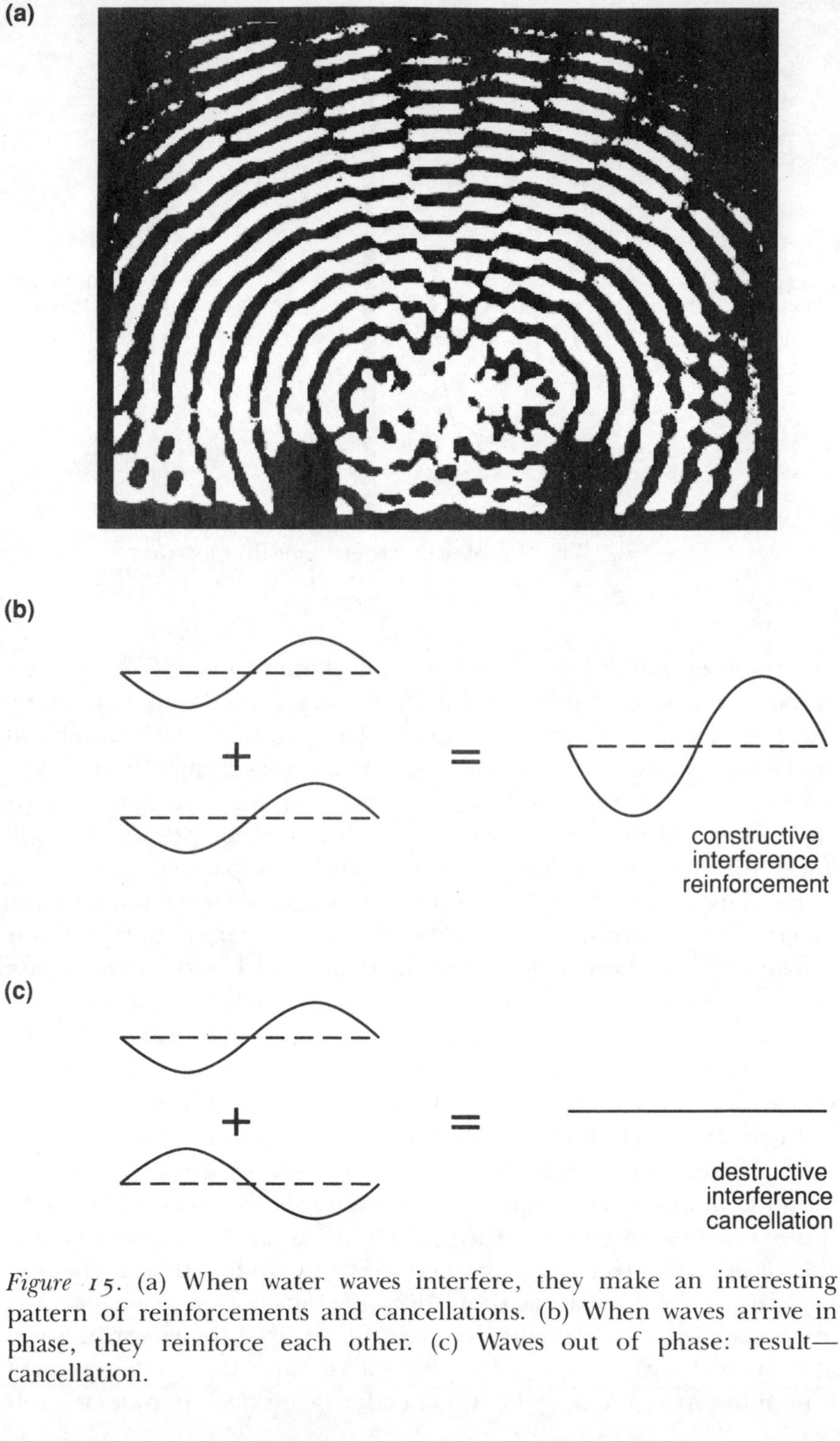

Figure 15. (a) When water waves interfere, they make an interesting pattern of reinforcements and cancellations. (b) When waves arrive in phase, they reinforce each other. (c) Waves out of phase: result—cancellation.

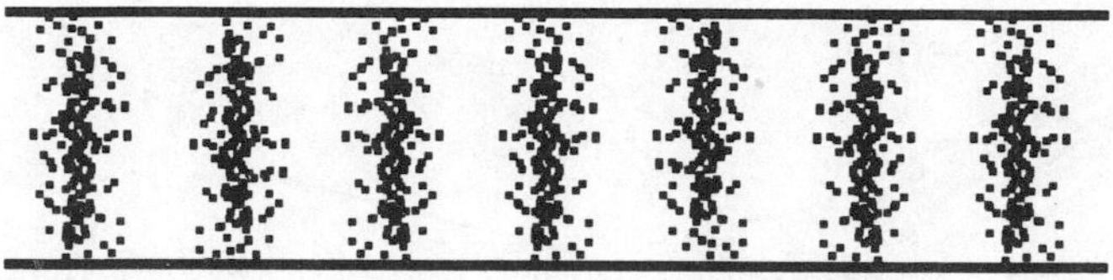

Figure 16. The interference pattern of flashes on screen.

is the totality of spots made by a large number of electrons that looks like the wave interference pattern.

Suppose we take an intellectual risk and make the electron beam very weak—so weak that at any one moment only one electron arrives at the slits. Do we still get an interference pattern? Quantum mechanics unambiguously says yes. We cannot, you may object, get interference without a split beam. Doesn't it take two waves to interfere? Can a single electron split, pass through both slits, and interfere with itself? Yes, it can. Quantum mechanics says yes to all these questions. As Paul Dirac, one of the pioneers of the new physics, put it: "Each photon [or here, electron] interferes only with itself." The proof that quantum mechanics offers for this preposterous proposition is mathematical, but this one proposition is responsible for all the miraculous magic that quantum systems are capable of and that has been verified by myriad experiments and technologies.

Try to imagine that an electron is passing 50 percent through one slit and 50 percent through the other slit. It is easy to get exasperated and to disbelieve this strange consequence of quantum mathematics. Does the electron really pass through both slits at the same time? Why should we take that for granted? We can find out by looking. We can focus a flashlight (metaphorically speaking) on a slit to see which hole the electron is really passing through.

So we turn the light on, and as we see an electron passing through a particular slit, we look to see where the flash appears on the fluorescent screen (fig. 17). What we find is that every time an electron goes through a slit, its flash appears just behind the slit it passes through. The interference pattern has disappeared.

What is happening in this experiment can be understood, in the first place, as a play of the uncertainty principle. As soon as we locate the electron and determine which slit it passes through, we lose the information about the electron's momentum. Electrons are very delicate; the collision with the photon that we are using to observe it affects it so that its momentum changes by an unpredictable amount.

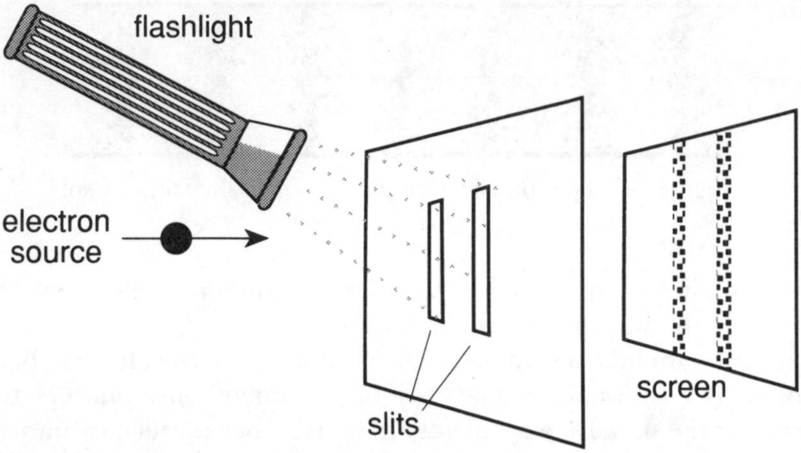

Figure 17. When we try to identify which slit the electron passes through by focusing a flashlight on the slits, the electron shows its particle nature. There are only two fringes—exactly what we would expect if the electrons were miniature baseballs.

The electron's momentum and the wavelength are related: This is de Broglie's great discovery that quantum mathematics incorporates. Thus losing the information about the electron's momentum is the same as losing information about its wavelength. If there were interference fringes, we would be able to measure the wavelength from their spacing. The uncertainty principle says that as soon as we determine which slit the electron is passing through, the process of looking destroys the interference pattern.

You must realize that the measurements on the electron's position and momentum are really complementary, mutually exclusive processes. We can concentrate on the momentum and measure the wavelength—and thus the momentum—of the electron from the interference pattern, but then we cannot tell which slit the electron goes through. Or we can concentrate on the position and lose the interference pattern, the information about the wavelength and momentum.

There is a second, even more subtle way to understand and reconcile all this—the way of the complementarity principle. Depending on which apparatus we choose, we see the particle aspect (for example, with a flashlight) or the wave aspect (no flashlight).

To understand the complementarity principle as saying that quantum objects are both wave and particle but that we can see only one attribute with a particular experimental arrangement is certainly correct, but our experience is teaching us some subtleties. For example, we must also say that the electron is neither a wave (because the wave aspect never manifests for a single electron) nor a particle (because it appears on the screen at places forbidden for particles). Then, if we are cautious in our logic, we must also say that the photon is neither not-wave nor not-particle, just so there is no misunderstanding about our use of the words *wave* and *particle*. This is much like the logic of the idealist philosopher Nagarjuna in the first century A.D., the most astute logician of the *Mahayana* Buddhist tradition.[2] Eastern philosophers communicate their understanding of ultimate reality as *neti, neti* (not this, not that). Nagarjuna formulated this teaching into four negations:

It does not exist.
It does not not exist.
It does not both exist and not exist.
Nor does it neither exist nor not exist.

To understand complementarity more clearly, suppose we go back to the previous experiment, this time using weak batteries to make the flashlight that we shine on the electrons somewhat dimmer. When we repeat the experiment of figure 17 with dimmer and dimmer flashlights, we find that some of the interference pattern begins to reappear, becoming more and more prominent as we make the flashlight dimmer and dimmer (fig. 18). When the flashlight is turned off completely, the full interference pattern comes back.

As the flashlight dims, the number of photons scattering off the electrons decreases, so some of the electrons entirely escape being "seen" by the light. Those electrons that are seen appear behind slit 1 or slit 2, just where we would expect them. Each of the unseen electrons splits and interferes with itself to make the wave-interference pattern on the screen when enough electrons have arrived there. In the limit of strong light, only the particle nature of the electrons is seen; in the limit of no light, only the wave nature is

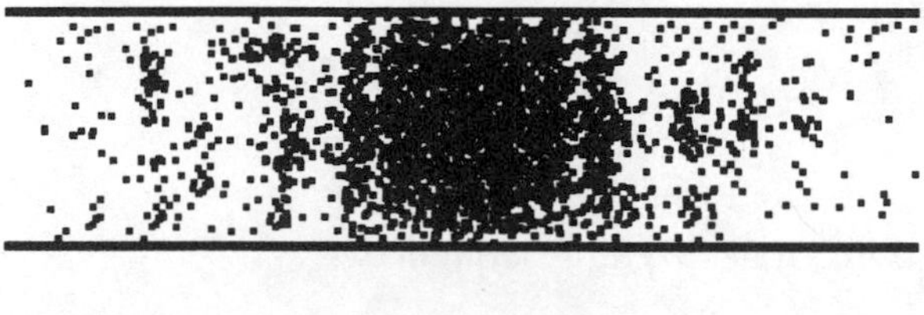

Figure 18. With a dimmer flashlight, some of the interference pattern returns.

seen. In the case of various intermediate situations of dim light, both aspects show up to a similarly intermediate degree: that is, here we are seeing electrons (though never the same electron) as simultaneously wave and particle. Thus the wave nature of a wavicle is not a property of the whole ensemble but must hold for each individual wavicle whenever we are not looking. That must mean that the wave aspect of a single quantum object is transcendent, since we never see it manifest.

A series of pictures helps explain what is going on (fig. 19). In the picture on the lower left, we see the letter W only; this corresponds to using a strong flashlight, which shows only the particle nature of the electrons. Then as we scan the ascending pictures, we begin to see the eagle—just as when we begin to make the light dimmer,

Figure 19. The W-Eagle sequence.

some electrons escape observation (and localization), and we start seeing their wave nature. Finally, in the last figure, upper right, only the eagle can be seen; the flashlight has been turned off, and the electrons are all waves now.

Niels Bohr once said: "Those who are not shocked when they first come across quantum theory cannot possibly have understood it." That shock yields to understanding as we begin to comprehend the play of the complementarity principle. The formal cadence of predictive science that holds for either wave or particle is transformed into a creative dance of a transcendent wavicle. When we localize the electron by finding out which slit it goes through, we reveal its particle aspect. When we do not localize the electron, ignoring which slit it goes through, we reveal its wave aspect. In the latter case, the electron passes through both slits.

The Delayed-Choice Experiment

Let us be clear about this unique characteristic of the complementarity principle: What attribute the quantum wavicle reveals depends on how we choose to observe it. Nowhere is the importance of conscious choice in the shaping of manifest reality better demonstrated than in the delayed-choice experiment suggested by physicist John Wheeler.

Figure 20 shows an apparatus in which a beam of light is split into two beams, each of equal intensity—one reflected and one transmitted—by using a half-silvered mirror M_1. These two beams are then reflected by two regular mirrors A and B to a crossing point P on the right.

To detect the wave aspect of the wavicle, we take advantage of the phenomenon of wave interference and put a second half-silvered mirror M_2 at P (fig. 20, bottom left). The two waves created by beam splitting at M_1 are now forced by M_2 to interfere constructively on one side of P (where if we place a photon counter, the counter ticks) and destructively on the other side (where a counter never ticks). Notice that when we are detecting the wave mode of the photons, we must agree that each photon splits up at M_1 and travels by both routes A and B, otherwise how can there be interference?

So when the mirror M_1 splits the beam, each photon potentially is ready to travel both paths. If we now choose to detect the particle mode of the photon wavicles, we take away the mirror M_2 at P (to

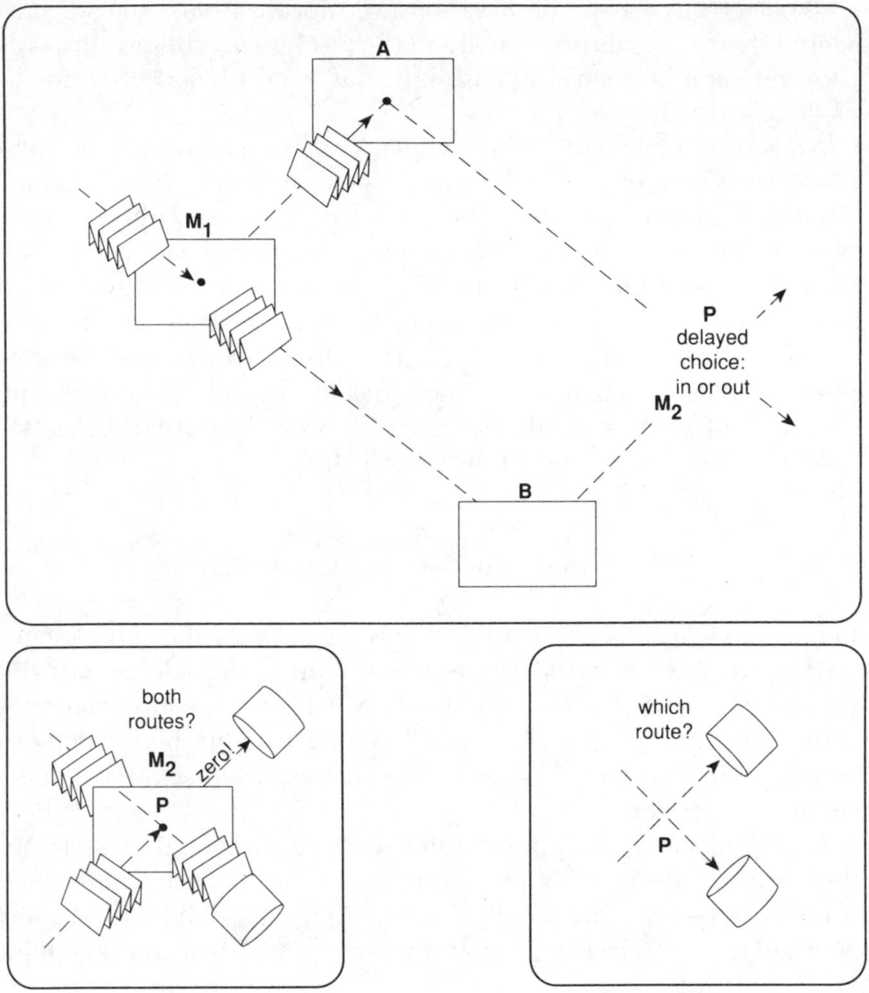

Figure 20. The delayed-choice experiment. LOWER LEFT: the arrangement for seeing the wave nature of photons. One of the detectors never detects any photons, signifying cancellation due to wave interference. The photon must have split and traveled both routes at the same time. LOWER RIGHT: the arrangement for seeing the particle nature of photons. Both detectors click, although only one at a time—signifying which route the photon takes.

prevent recombination and interference) and put counters past the point of crossing *P*, as shown in the lower right in figure 20. One or the other counter will tick, defining the localized path of a wavicle, the reflected path *A* or the transmitted path *B*, to show its particle aspect.

The subtlest aspect of the experiment is as follows: In the delayed-choice experiment, the experimenter decides at the very last moment, in the very last pico (10^{-12}) second (this has been done in the laboratory),[3] whether or not to insert the half-silvered mirror at *P*, whether or not to measure the wave aspect. In effect, this means that the photons have already traveled past the point of splitting (if you think of them as classical objects). Even so, inserting the mirror at *P* always shows the wave aspect and not inserting the mirror shows the particle aspect. Was each photon moving in one path or two? The photons seem to respond even to our delayed choice instantly and retroactively. A photon travels one path or both paths, exactly in harmony with our choice. How does it know? Is the effect of our choice preceding its cause in time? Says Wheeler: "Nature at the quantum level is not a machine that goes its inexorable way. Instead, what answer we get depends on the question we put, the experiment we arrange, the registering device we choose. We are inescapably involved in bringing about that which appears to be happening."[4]

There is no manifest photon until we see it, and thus how we see it determines its attributes. Before our observation, the photon splits into two wave packets (a packet for each path), but these packets are only packets of possibilities for the photon; there is no actuality in space-time, no decision making at M_1. Does the effect precede its cause and violate causality? It certainly does—if you think of the photon as a classical particle always manifest in space-time. The photon, however, is not a classical particle.

From the viewpoint of quantum physics, if we put a second mirror at *P* in our delayed-choice experiment, the two split-up packets in potentia combine and interfere; there is no problem. If there were a mirror at *P* and we removed it at the last-possible pico second, detecting the photon in path *A*, say, it would seem that the photon is responding to our delayed choice retroactively by traveling only in one path. In this case, therefore, the effect seems to be preceding the cause. This result does not violate causality. How so?

You must comprehend a more subtle way of looking at the second particle-aspect detection experiment, as elucidated by Heisenberg: "If now an experiment yields the result that the photon is, say, in the reflected part of the [wave] packet [path *A*], then the probability of finding the photon in the other part of the packet immediately becomes zero. The experiment in the position of the reflected packet then exerts a kind of action . . . at the distant point occupied by the transmitted packet, and one sees [that] this action is propagated with a velocity greater than light. However, it is also obvious that this kind of action can never be utilized to transmit a signal so that it [does] . . . not conflict with the postulates of the theory of relativity."[5]

This action-at-a-distance is an important aspect of the collapse of the wave packet. The technical term that we use for such action-at-a-distance is *nonlocality*—action transmitted without signals that propagate through space. Signals that propagate through space, taking a finite time because of the Einsteinian speed limit, are called *local signals*. So the collapse of the quantum wave is nonlocal.

Note that the point Heisenberg makes holds with or without delayed choice. In the quantum view, the critical point is that we choose the specific outcome that manifests; when, in time, we choose that outcome is unimportant. The wave splits whenever there are two available paths, but the split occurs only in potentia. When, later, we observe the photon in one path because we so choose (by removing the mirror at *P*), our collapsing the wave in one path exerts a nonlocal influence on the wave in the other path that negates the possibility of the photon being seen in that other path. Such a nonlocal influence may seem retroactive, but we are influencing only possibilities in potentia; there is no breakdown of causality because, as Heisenberg says, we cannot transmit a signal through this kind of device.

In our search for the meaning and structure of reality, we are facing the same puzzle that confronted Winnie-the-Pooh:

> 'Hallo!' said Piglet, 'what are *you* doing?'
> 'Hunting,' said Pooh.
> 'Hunting what?'
> 'Tracking something,' said Winnie-the-Pooh very mysteriously.
> 'Tracking what?' said Piglet, coming closer.
> 'That's just what I ask myself. I ask myself, What?'
> 'What do you think you'll answer?'
> 'I shall have to wait until I catch up with it,' said Winnie-the-Pooh. 'Now, look there.' He pointed to the ground in front of him. 'What do you see there?'
> 'Tracks,' said Piglet. 'Paw-marks.' He gave a little squeak of excitement. 'Oh, Pooh! Do you think it's a——a——a Woozle?'
> 'It may be,' said Pooh. 'Sometimes it is, and sometimes it isn't. You never can tell with paw-marks.'
> 'Wait a moment,' said Winnie-the-Pooh, holding up his paw. He sat

down and thought, in the most thoughtful way he could think. Then he fitted his paw into one of the Tracks . . . and then he scratched his nose twice, and stood up.

'Yes,' said Winnie-the-Pooh. 'I see now,' said Winnie-the-Pooh. 'I have been Foolish and Deluded,' said he, 'and I am a Bear of No Brain at All.'

'You're the Best Bear in All the World,' said Christopher Robin soothingly.[6]

How puzzling indeed that the "woozle" tracks that the electron and other submicroscopic particles leave in our cloud chambers are, according to the new physics, merely extensions of ourselves.

The classical scientist looked at the world and saw his single vision of separateness. A couple of centuries ago, the English romantic poet William Blake wrote:

may God us keep
From single vision and Newton's sleep.[7]

Quantum physics is the answer to Blake's prayer. The quantum scientist who has learned the lesson of the complementarity principle knows better than to heed (apparent) separateness.

Quantum measurements interject our consciousness into the arena of the so-called objective world. There is no paradox in the delayed-choice experiment if we give up the idea that there is a fixed and independent material world even when we are not observing it. Ultimately, it boils down to what you, the observer, want to see. This reminds me of a Zen story.

Two monks were arguing about the motion of a flag in the wind. Said one: "The flag is moving." "No, the wind is moving," said the other. A third monk, who was passing by, made an observation that Wheeler would approve. "The flag is not moving. The wind is not moving. Your mind is moving."

Chapter 6

THE NINE LIVES OF SCHRÖDINGER'S CAT

MANY OF THE FOUNDERS of quantum physics had a hard time accepting its strange consequences. Schrödinger himself expressed his reservations about the probability-wave interpretation of quantum mechanics in the paradox now known as Schrödinger's cat.

Suppose that we put a cat in a cage with a radioactive atom and a Geiger counter. The radioactive atom will decay in accordance with probabilistic rules. If the atom decays, the Geiger counter will tick, the ticking will trigger a hammer, the hammer will break a bottle of poison, and the poison will kill the cat. Let us suppose that there is a 50 percent chance of this occurring within an hour (fig. 21).

How, then, would quantum mechanics describe the state of the cat after an hour? Of course, if we look, we will find the cat to be either alive or dead. What if we do not look? The probability that the cat is dead is 50 percent. The probability that the cat is alive is also 50 percent.

If you think classically, in the manner of the material realists, and take determinism and causal continuity as your guiding principles, then you might make a mental analogy to the situation in which someone has flipped a coin and then has hidden it under his palm. You do not know whether the outcome is heads or tails, but of course, it is one or the other. The cat is either dead or alive, with a 50 percent chance for each outcome. You just do not know which

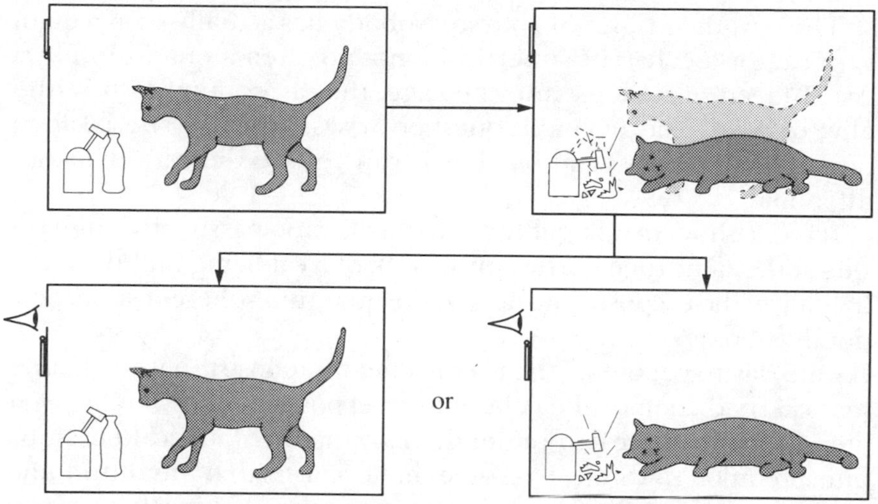

Figure 21. The paradox of Schrödinger's cat. After an hour with a radioactive atom in a cage, the cat becomes a coherent superposition of a half-dead and half-alive cat. Observation always reveals either a dead cat or a live cat. (Reprinted from A. Goswami, *Quantum Mechanics*; permission granted by Wm. C. Brown, Inc., publisher.)

outcome has, in fact, been realized. This scenario is not what the mathematics of quantum mechanics portrays. Quantum mechanics deals with probabilities very differently. It describes the state of the cat at the end of the hour as half alive and half dead. Inside the box is, quite literally, "a coherent superposition of a half-alive and a half-dead cat," to use the proper jargon. The paradox of a cat that is dead and alive at the same time is a consequence of the way in which we do our calculations in quantum mechanics. However bizarre its consequences, we must take this mathematics seriously because the same mathematics gives us the marvels of transistors and lasers.

The following parody of T. S. Eliot's *Old Possum's Book of Practical Cats* summarizes this absurd situation:

> Schrödinger's cat's a mystery cat,
> he illustrates the laws;
> the complicated things he does
> have no apparent cause;
> he baffles the determinist,
> and drives him to despair
> for when they try to pin him down—
> *the quantum cat's not there!*[1]

The parody is right, of course. Nobody has actually seen a quantum cat, or a coherent superposition—not even a quantum physicist. Indeed, if we look into the cage, the cat is found to be either alive or dead. The inevitable question arises: What's so special about our making an observation that it can resolve the cat's diabolical dilemma?

It is one thing to talk glibly about an electron passing through two slits at the same time, but when we talk of a cat being half dead and half alive, the preposterousness of the quantum coherent superposition hits home.

One way to get out of the predicament is to insist that the mathematical prediction of the coherent superposition must not be taken literally. Instead, we can pretend, following the statistical-ensemble interpretation favored by some materialists that quantum mechanics makes predictions only about experiments involving a very large number of objects. If there were ten billion cats, all in individual cages set up identically, quantum mechanics would tell us that half of them would be dead in an hour, and surely observation would bear out the truth of that assertion. Maybe for a single cat the theory just does not apply. In the last chapter a similar argument was made for electrons. It is a fact, however, that the ensemble interpretation encounters difficulty explaining even the simple double-slit interference pattern.[2]

Furthermore, the ensemble interpretation is tantamount to giving up quantum mechanics as a physical theory for the description of a single object or of a single event. Since single events do occur (even single electrons have been isolated), we must be able to talk about single quantum objects. Indeed, quantum mechanics was formulated to apply to single objects, notwithstanding the paradoxes that it raises by doing so. We must face up to Schrödinger's paradox and seek a way to resolve it. The alternative is to have no physics at all for single objects—a wholly undesirable alternative.

Many physicists today hide behind the anti-metaphysical philosophy of logical positivism when dealing with the paradox of Schrödinger's cat. Logical positivism is the philosophy that grew out of the Viennese philosopher Ludwig Wittgenstein's *Tractatus Logico-Philosophicus*, a work in which he argued, famously, that "Whereof one cannot speak, thereof one must remain silent." Following this dictum, these physicists—we may call them neo-Copenhagenists—maintain that we should confine our discussion of reality to what is seen instead of trying to assert the reality of something that we cannot observe. For them, the point is that we never see the coherent superposition. Is the unobserved cat half-dead and half-alive? You cannot ask that question, they would say, because it cannot be answered. This, of course, is sophistry. A question that cannot be answered directly can nonetheless be approached circuitously, and its answer can be calculated on the grounds of consistency with what we can directly know. Moreover, avoiding metaphysical questions entirely is not consistent with the spirit of the original Copenhagen interpretation and the way in which Bohr and Heisenberg saw things.

The Copenhagen interpretation, if one follows Bohr, lessens the absurdity of the half-dead, half-alive cat by means of the complementarity principle: The coherent superposition is an abstraction; as an abstraction, the cat is able to exist as both live and dead. This is a complementary description, complementary to the dead or alive description that we give when we do see the cat. According to Heisenberg, the coherent superposition—the half-dead, half-alive cat—exists in transcendent potentia. It is our observation that collapses the cat's dichotomous state into a single one.

What sense are we to make of this notion of a half-dead, half-alive cat existing in potentia? An answer that sounds like science fiction has come from the physicists Hugh Everett and John Wheeler.[3] According to Everett and Wheeler, both possibilities, live cat and dead cat, occur—but in different realities, or parallel universes. For every live cat we find in the cage, prototypes of us in a parallel universe open a prototype cage only to discover a prototype cat that is dead. An observation of the cat's dichotomous state forces the universe itself to split into parallel branches. This is an intriguing idea, and some science fiction writers (notably Philip K. Dick) make good use of it. Unfortunately, this is also a costly idea. It would double the amount of matter and energy each time an observation forces the universe to bifurcate. It offends our taste for parsimony, which may be a prejudice but is nonetheless a cornerstone of scientific reasoning. Furthermore, since the parallel universes do not interact, this interpretation is difficult to put to experimental test and therefore not useful from a scientific point of view. (Fiction is more tractable. In Philip Dick's *The Man in the High Castle*, the parallel universes do interact. How else would there be a story?)

Fortunately, an idealist resolution presents itself: Since our observation magically resolves the dichotomy of the cat, it must be us—our consciousness—that collapses the cat's wave function. Material realists do not like this idea, because it makes consciousness an independent, causal entity; admitting that would be like putting nails in the coffin of material realism. Materialism notwithstanding, such luminaries as John von Neumann, Fritz London, Edmond Bauer, and Eugene Paul Wigner have endorsed this resolution to the paradox.[4]

The Idealist Resolution

In the idealist resolution, it is observation by a conscious mind that resolves the alive-or-dead dichotomy. Like Platonic archetypes, coherent superpositions exist in the never-never land of a transcendent order until we collapse them, bringing them into the world of manifestation with an act of observation. In the process, we choose one facet out of two, or many, that are permitted by the Schrödinger equation; it is a limited choice, to be sure, subject to the overall probability constraint of quantum mathematics, but it is a choice nevertheless.

Even if material realism is false, should we hastily give up scientific objectivity and invite consciousness into our science? Paul Dirac, one of the pioneers of quantum physics, once said that great breakthroughs in physics always involve giving up some great prejudice. Perhaps the time has come to give up the prejudice of strong objectivity. Bernard d'Espagnat suggests that the objectivity permitted by quantum mechanics is weak objectivity.[5] Instead of the observer-independence of events demanded by strong objectivity, quantum mechanics allows a certain meddling by the observer—but in such a way that the interpretation of the events does not depend on any particular observer. Thus weak objectivity is observer-invariance of events: Irrespective of who the observer is, the event remains the same. In view of the subjective choice involved in individual measurements, it is a statistical principle to be sure, and observer invariance holds only for a large number of observations, which is nothing new. Having long accepted the probability interpretation of quantum mechanics, we are already committed to accepting the statistical nature of some of our scientific principles: the causality principle, for example. As cognitive psychology routinely demonstrates, we can certainly do science with weak objectivity defined in this way. We do not really need strong objectivity.

The consciousness resolution of Schrödinger's paradox is the most straightforward one—so much so that it is sometimes referred to as the naive resolution. Many questions have been raised about this resolution, however, and only by answering these questions can we overcome the accusation of naiveté.

Questions About the Idealist Resolution

One question you may still be asking is, How can a cat be half-dead and half-alive? It cannot, if you are thinking as a material realist. The material realist must assume that the state of the cat at every moment is either this or that, dead or alive, in a causally continuous fashion. Materialist thinking, however, is the result of assumptions of causal continuity and an either/or description of events. These assumptions are not necessarily true, especially when they are tested against quantum mechanical experiments.

To an idealist philosopher the paradox of a cat being both dead and alive is not particularly disturbing. In a Zen story, a master was shown a so-called dead man whose funeral was being prepared. When he was asked if the man was dead or alive, the Zen master replied, "I cannot say." How could he? According to idealism, the essence of a man, consciousness, never dies. So it would be incorrect to say outright that the man is dead. When a man's body is being prepared for his funeral, however, it would be ridiculous to say that the man is alive.

Is the cat dead or alive? Zen master Joshu answered the question Does a dog have Buddha nature? by replying, "mu" (pronounced moo). Again, to say no would be wrong since all creatures, according to Buddha's teaching, have Buddha nature. To say yes would also be tricky because the Buddha nature is to be realized and lived—not a matter of intellectual truth. So the answer is mu: neither yes nor no.

Quantum mechanics seems to imply an idealist philosophy like that of the Zen masters when it asserts that Schrödinger's cat is, at the end of an hour, half dead and half alive. How can this be? How can consciousness be decisive in shaping the reality of the physical world? Does this not imply the primacy of consciousness over matter?

If Schrödinger's cat is both alive and dead before we look inside the box but has a unique state (alive or dead) after we look, then we must be doing something just by looking. How can a tiny peek have an effect on the physical state of a cat? These are questions that realists ask when trying to refute the idea that the coherent superposition is collapsed by consciousness.

Yes, the idealist resolution does imply the action of consciousness upon matter. That action, however, poses a problem only for material realism. In this philosophy, consciousness is an epiphenomenon of matter, and it seems impossible that an epiphenomenon of matter could act on the very fabric of which it is built—in effect causing itself. That causal paradox is avoided by monistic idealism, in which consciousness is primary. In consciousness, coherent superpositions are transcendent objects. They are brought into immanence only when consciousness, by the process of observation, chooses one of the many facets of the coherent superposition, though its choice is constrained by the probabilities allowed by the quantum calculus. (Consciousness is lawful. The creativity of the cosmos comes from the creativity of its quantum laws, not from arbitrary lawlessness.)

According to monistic idealism, objects are already in consciousness as primordial, transcendent, archetypal possibility forms. The collapse consists not of doing something to objects via observing but of choosing and of recognizing the result of that choice.

Look back once more at the gestalt illustration "My Wife and My Mother-in-Law" (fig. 12). In this illustration, two pictures are superimposed. When we see the wife (or the mother-in-law), we are not doing anything to the picture. We are simply choosing and recognizing our choice. The process of collapse by consciousness is something like this.

There are, however, dualists who try to explain the action of consciousness in Schrödinger's paradox by finding evidence of psychokinesis: the ability to move matter with the mind.[6] Eugene Paul Wigner argues that if a quantum object can affect our consciousness, then our consciousness must be able to affect a quantum object. The evidence for psychokinesis, however, is scanty and dubious. Furthermore, evidence from another paradox—that of Wigner's friend—effectively rules out a dualistic interpretation.

THE PARADOX OF WIGNER'S FRIEND

Suppose that two people simultaneously open the cage of the cat. If the observer chooses the outcome of collapse, as idealism seems to imply, then suppose the two observers chose differently, would that not create a problem? If we say no, only one of the observers gets to choose, the realist is not satisfied and rightly so.

The paradox of Wigner's friend, formulated by physicist Eugene Wigner, goes something like this: Suppose that instead of observing the cat himself, Wigner asks a friend to do so. His friend opens the cage, sees the cat, and then reports the results of his observation to Wigner. At this point, we can say that Wigner has just actualized the reality that includes his friend and the cat. There is a paradox here: Was the cat alive or dead when Wigner's friend observed it but before he reported the observation? To say that the state of the cat did not collapse when his friend observed the cat is to maintain that his friend remained in a state of suspended animation until Wigner asked him—that his friend's consciousness could not decide whether the cat was alive or dead without Wigner's prodding. That sounds a lot like solipsism—the philosophy that posits you as the only conscious being with everybody else imaginary. Why should Wigner be the privileged one who gets to collapse the cat's state function?

Suppose we say, instead, that Wigner's friend's consciousness collapses the superposition. Does that not open up a hornet's nest? If Wigner and his friend look at the cat simultaneously, whose choice is going to count? What if the two observers choose differently? The world would be pandemonium if individual people were to decide the behavior of the objective world, because we know subjective impressions are often contradictory. The situation in such a case would be like that of people coming from different directions and choosing the color (red or green) of a traffic light at will. This argument is often regarded as a fatal blow against the consciousness resolution of Schrödinger's paradox. It is fatal, however, only to a dualist interpretation. Let us explore Wigner's paradox in more detail to see why this is so.

Wigner has compared his paradoxical state of affairs with one in which an inanimate apparatus is used to make the observation. When a machine is used, there is no paradox. There is nothing paradoxical or upsetting about a machine being in limbo for a while, but experience says that there is something decisive about a conscious being's observation. As soon as a conscious being observes, the material reality becomes manifest in a unique state. Says Wigner:

> It follows that the being with a consciousness must have a different role in quantum mechanics than the inanimate measuring device. . . . This argument implies that 'my friend' has the same types of impressions and sensations as I—in particular, that, after interacting with the object, he is not in that state of suspended animation. . . . It is not necessary to see a contradiction here from the point of view of orthodox quantum mechanics, and there is none if we believe that the alternative is meaningless, whether my friend's consciousness contains . . . the impression of having seen [either a dead cat or a live cat]. However, to deny the existence of the consciousness of a friend to this extent is surely an unnatural attitude, approaching solipsism, and few people in their heart will go along with it.[7]

The paradox is subtle, but Wigner is right. We do not have to say that until Wigner manifests his friend, his friend stays in a state of suspended animation. Nor do we have to resort to solipsism. There is an alternative.

Wigner's paradox arises only when he makes the unwarranted dualist assumption that his consciousness is separate from his friend's. The paradox disappears if there is only one subject, not separate subjects as we normally understand them. The alternative to solipsism is a unitive subject-consciousness.

When I observe, what I see is the whole world of manifestation, but this is not solipsism, because there is no individual I that sees as opposed to other I's. Erwin Schrödinger was right when he said: "Consciousness is a singular for which there is no plural." Etymology and orthography have preserved the singularity of consciousness. The existence in language of such terms as *I* and *my*, however, leads us into a dualistic trap. We think of ourselves as separate because we speak of ourselves in that way.

Similarly, people fall into thinking about having consciousness, as in the question, Does a cat have consciousness? It is only in material realism that consciousness is something merely to be possessed. Such a consciousness would be determined, not free, and would not be worth having.

THE WATCHED POT DOES BOIL

Consider another wrinkle in Schrödinger's paradox. Suppose that Schrödinger's cat is itself a conscious being. The concept becomes even more acute by assuming a human being inside the cage with the radioactive atom, the bottle of poison, and all the rest. Suppose, then, that we open the cage after an hour and, if he is still alive, ask him if he experienced a half-alive, half-dead state? Nope! he will say. Are we getting into trouble here for the idealist interpretation? Consider for a moment. What if we ask him, instead, whether he experienced being alive all the time. After some reflection, if ours is a reflective subject, he will probably say no. You see, we are not conscious of our bodies all the time. In fact, we have very little consciousness of our bodies under ordinary circumstances. So here is what the idealist interpretation may describe as happening. During the hour, every now and then, he was conscious that he was alive. In other words, he regarded himself. At those times his wave function collapsed, and fortunately the choice was the alive state each time. In between these moments of wave collapse his wave function expanded and became a coherent superposition of dead and alive in the transcendent domain that is beyond experience.

You know how we see a motion picture. Our brain-mind cannot discern the individual still pictures that race before our eyes at a speed of twenty-four frames per second. Similarly, what seems to be continuity to a human observer watching himself is really a mirage consisting of many discontinuous collapses.

This last argument also means that we cannot save Schrödinger's cat from the diabolical result of the decay of the radioactive atom by constantly looking at it, and thus somehow collapsing its wave function continuously and keeping it alive. It is a noble thought, but it will not work—for the same reason that a watched pot boils, even though the adage suggests otherwise. It is a good thing, too, that the watched pot boils, because if we could prevent change just by staring at an object, the world would be full of narcissists trying to escape aging and death by meditating on themselves.

Heed Erwin Schrödinger's reminder: "Observations are to be regarded as discrete, discontinuous events. Between there are gaps which we cannot fill in."

The resolution of the Schrödinger's cat paradox tells us a great deal about the nature of consciousness. It chooses among alternatives when it manifests the material reality; it is transcendent and unitive; and its doings elude our normal mundane perception. Admittedly, none of these aspects of consciousness is self-evident to common sense. Try to suspend your disbelief and remember what Robert Oppenheimer once said: "Science is uncommon sense."

Quantum collapse is a process of choosing and recognizing by a conscious observer; there is ultimately only one observer. This means that we have one other classic paradox to resolve.

WHEN IS A MEASUREMENT COMPLETE?

To some realists a measurement is complete when a classical measuring apparatus, such as the Geiger counter in Schrödinger's cat cage, measures a quantum object; it is complete when the counter ticks. Note that if we accept such a solution, the paradox of the cat's dichotomous state does not arise.

This reminds me of a story. Two elderly gentlemen were talking, and one was complaining about his chronic gout. The other said with some pride: "I never have to worry about gout; I take a cold shower every morning." The gentleman with gout looked at him quizzically and replied: "So you got chronic cold shower instead!"

These realists try to replace the dichotomy of Schrödinger's cat with another: a classical-quantum dichotomy. They divide up the world into quantum objects and their classical measurement apparatuses. Such a dichotomy, however, cannot be upheld; neither is it needed. We can assert that all objects obey quantum physics (the unity of physics!) and yet answer satisfactorily the question, When is a measurement complete?

What defines a measurement? Put slightly differently, when can we say that a quantum measurement is completed? We can approach the answer historically.

Werner Heisenberg, who proposed the uncertainty principle, formulated a thought experiment that Bohr clarified further. Recently David Bohm has given an account of the experiment, and I will adapt it here.[8] Suppose a particle is at rest in the target plane of a microscope and that we are analyzing its observation in terms of classical physics. To observe the target particle, we focus (with the help of the microscope) another particle that is deflected by the target particle onto a photographic emulsion plate, leaving a track. Based on the track and on our knowledge of how the microscope works, we can determine, according to classical physics, both the position of the target particle and the momentum imparted to it at the moment of deflection. The specific experimental conditions do not influence the final result.

All this changes in quantum mechanics. If the target particle is an atom and if we are looking at it through an electron microscope in which an electron is deflected from the atom onto a photographic plate (fig. 22), the following four considerations enter:

1. The deflected electron must be described as both a wave (while it is traveling from the object *O* to the image *P*) and as a particle (at arrival at *P* and while leaving the track *T*).
2. Because of this wave aspect of the electron, the image point *P* tells us only the probability distribution of the position of the object *O*. In other words, the position is determined only within a certain uncertainty Δx (pronounced delta ex).
3. Similarly, argued Heisenberg, the direction of the track *T* gives us only the probability distribution of the momentum of *O* and thus determines the momentum only within an uncertainty Δp (Delta pee). Using simple mathematics, Heisenberg was able to show that the product of the two uncertainties is equal to or greater than Planck's constant. This is Heisenberg's uncertainty principle.
4. In a more detailed mathematical account, Bohr pointed out that it is impossible to specify the wave function of the observed atom separately from that of the electron that is used to see it. In truth, said Bohr, the wave function of the electron cannot be unentangled from that of the photographic emulsion. And so on. We cannot draw the line in this chain without ambiguity.

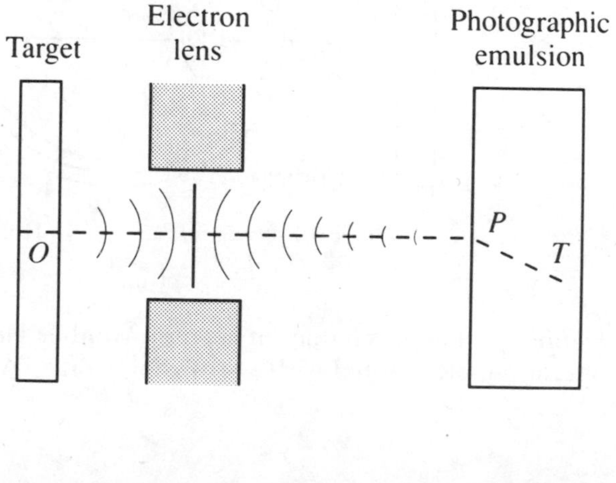

Figure 22. The Bohr-Heisenberg microscope. (Reprinted with permission from J. A. Schumacher.)

In spite of the ambiguity in drawing the line, Bohr felt that we must draw it because of the "indispensable use of classical concepts in the interpretation of all proper measurements." The experimental arrangement, said Bohr reluctantly, must be described in totally classical terms. The dichotomy of quantum waves must be assumed to terminate with the measuring apparatus.[9] As was pointed out cogently by the philosopher John Schumacher, however, all actual experiments have a second Heisenberg microscope built into them:[10] The process of seeing the emulsion track involves the same kind of consideration that led Heisenberg to the uncertainty principle (fig. 23). Photons from the emulsion track are amplified by an experimenter's own visual apparatus. Can we ignore the quantum mechanics of our own seeing? If not, is our brain-mind-consciousness not inexorably connected with the measurement process?

IS THE CAT QUANTUM OR CLASSICAL?

When you think about it, it becomes clear that Bohr was replacing one dichotomy, that of the cat, with another, that of a world divided into quantum and classical systems. According to Bohr, we cannot separate the wave function of the atom from the rest of the environ-

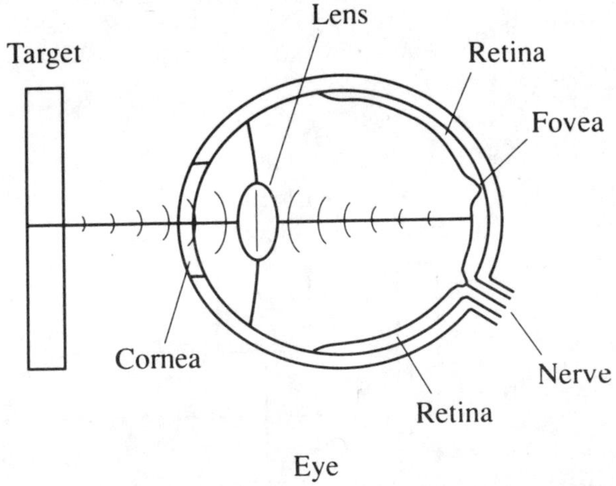

Figure 23. The mechanics of seeing. Another Heisenberg microscope in operation? (Reprinted with permission from J. A. Schumacher.)

ment in the cat's cage (the various measuring devices for the atom's decay, such as the Geiger counter, the poison bottle, and even the cat), and the line we draw between the micro world and the macro world is quite arbitrary. Unfortunately, Bohr also maintained that we must accept that the observation by a machine—a measuring apparatus—resolves the dichotomy of a quantum wave function.

Any macro body (the cat or any observing machine), however, is ultimately a quantum object; there is no such thing as a classical body unless we are willing to admit a vicious quantum/classical dichotomy in physics. It is true that a macro body's behavior can be predicted in most situations from the rules of classical mechanics. (Quantum mechanics gives the same mathematical predictions as does classical mechanics in such cases—this is the correspondence principle that Bohr himself pioneered.) For this reason we often loosely refer to macro bodies as being classical. The measurement process, however, is not such a case, and the correspondence principle does not apply to it. Bohr knew this, of course. In his celebrated debates with Einstein, he often invoked quantum mechanics for describing macro bodies of measurement in order to refute the acute objections that Einstein raised to probability waves and to the uncertainty principle.[11]

As an example of the debate between Bohr and Einstein, consider the double-slit arrangement but include an additional facet. Suppose that before their incidence on the double slit, the electrons pass through a single slit in a diaphragm—its purpose being the accurate definition of the starting point of the electrons. Einstein suggested that this initial slit be mounted on some extremely light springs (fig. 24). Einstein argued that if the first slit deflects an electron to the upper of the two slits, then the first diaphragm will recoil downward from the principle of conservation of momentum. The opposite would happen if an electron is deflected downward, toward the bottom slit. Thus the measurement of the recoil of the diaphragm will tell us which slit the electron really passes through, information that quantum mechanics is supposed to deny. If the first diaphragm is really classical, then Einstein is right. Defending quantum mechanics, Bohr pointed out that ultimately the diaphragm also obeys quantum uncertainty. Thus if its momentum is measured, its position becomes uncertain. This broadening of the first slit effectively wipes out the interference pattern, as Bohr was able to demonstrate.

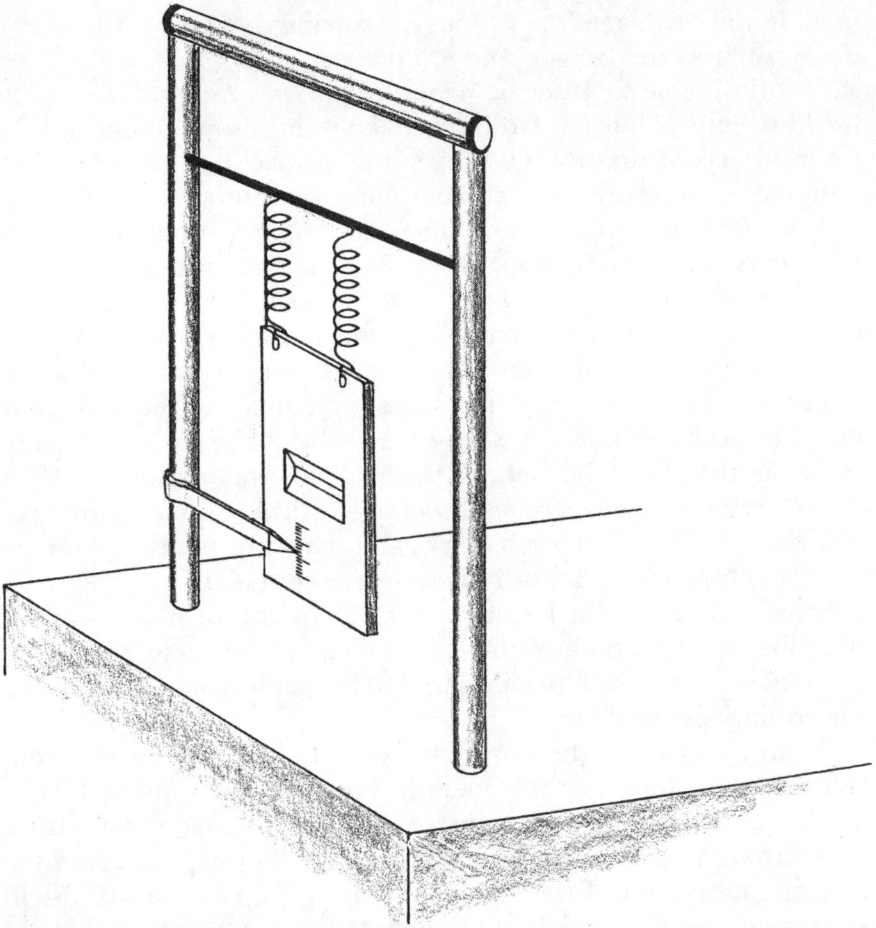

Figure 24. Einstein's spring-mounted initial slit for the double-slit experiment. If electrons go through a slit mounted on springs, as shown, before going through the double-slitted screen (not shown), is it possible to tell which slit an electron goes through without destroying the interference pattern?

Suppose further, however, that a complementarity principle is operating and that sometimes a macro apparatus does take on the quantum dichotomy (as shown by the Bohr-Einstein debate), but that at other times it does not—as happens with a measuring apparatus. This idea, called macrorealism, is ingenious, and it comes from the brilliant physicist Tony Leggett, whose work has inspired a beautiful experimental device called SQUID (Superconducting Quantum Interference Device).[12]

Ordinary conductors conduct electricity, but they always offer some resistance to the flow of electric current through them, which results in a loss of electrical energy as heat. In contrast, superconductors allow a current to flow without resistance. Once you set up a current through a superconducting loop, the current will flow, practically forever—even without a source of power. Superconductivity is due to a special correlation between electrons that extends over the whole body of the superconductor. It takes energy for the electrons to break away from this correlated state, thus the state is relatively immune to the random thermal motion present in an ordinary conductor.

The SQUID is a piece of superconductor with two holes in it that very nearly touch at a point called the weak link (fig. 25). Suppose we set up a current in the loop around one of the holes. A current sets up a magnetic field just as any electromagnet does, and the field lines representing the magnetic field pass through the hole—that, too, is usual. What is unusual for a superconductor is that the magnetic flux, the number of field lines per unit area, is quantized;

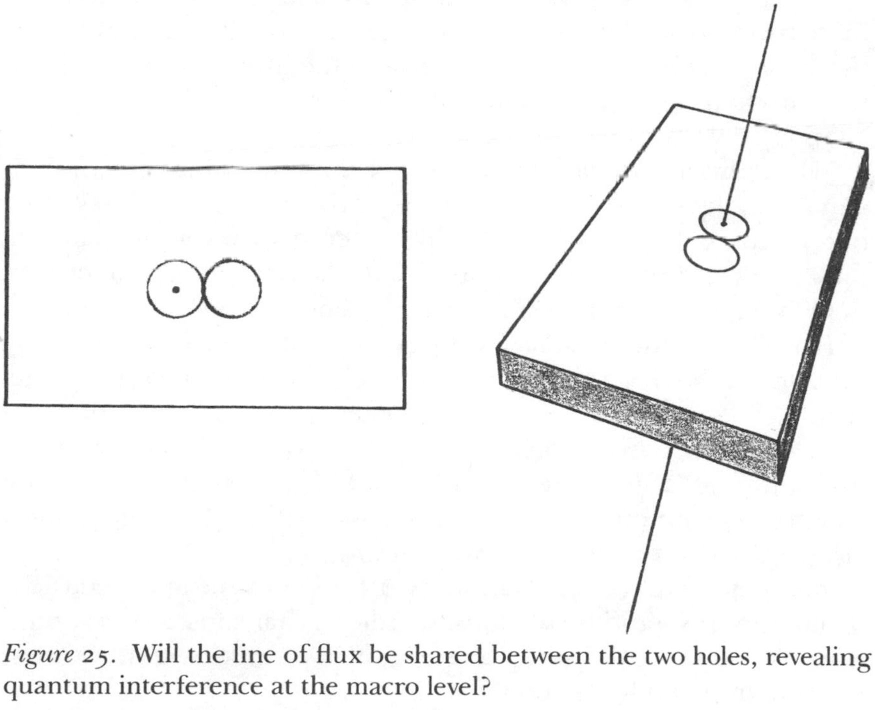

Figure 25. Will the line of flux be shared between the two holes, revealing quantum interference at the macro level?

the magnetic flux passing through the hole is discrete. This gave Leggett his key idea.

Suppose we set up such a small current that there is only one quantum of flux. Then we have created a double slit-type interference question. If there is only one hole, then obviously the flux quantum can be anywhere in it. If the link between the two holes is too thick, then the flux will be localized in only one hole. With just the right size of weak link, might we set up quantum interference such that the flux quantum is in both holes at the same time, nonlocalized? If so, quantum coherent superpositions clearly persist even at the scale of macrobodies. If no such nonlocalization is seen, then we may be able to conclude that macrobodies really are classical and do not permit coherent superpositions as their allowed states.

So far, there is no evidence of any breakdown of quantum mechanics with SQUID, but Leggett strongly expects quantum theory to break down. Said he at a recent conference: "But occasionally at night, when the full moon is bright, I do what in the physics community is the intellectual equivalent of turning into a werewolf: I question whether quantum mechanics is the complete and ultimate truth about the physical universe. . . . I am inclined to believe that at *some* point between the atom and the human brain it [quantum mechanics] not only may but *must* break down."[13]

Spoken like a true material realist!

Many physicists feel inclined to ask the same questions that inspire Leggett, so the research with SQUID continues. I suspect that one of these days it will turn up evidence in favor of quantum mechanics and will show that quantum coherent superpositions are demonstrably present even in macrobodies.

If we do not deny that all objects ultimately pick up quantum dichotomy, then, as von Neumann first argued, if a chain of material machines measures a quantum object in a coherent superposition, they all in turn pick up the dichotomy of the object, ad infinitum (fig. 26).[14] How do we get out of the logjam that the von Neumann chain creates? The answer is startling. *By jumping out of the system, out of the material order of reality.*

We know that an observation by a conscious observer ends the dichotomy. It should be obvious, therefore, that consciousness must work from outside the material world; in other words, consciousness must be transcendent—nonlocal.

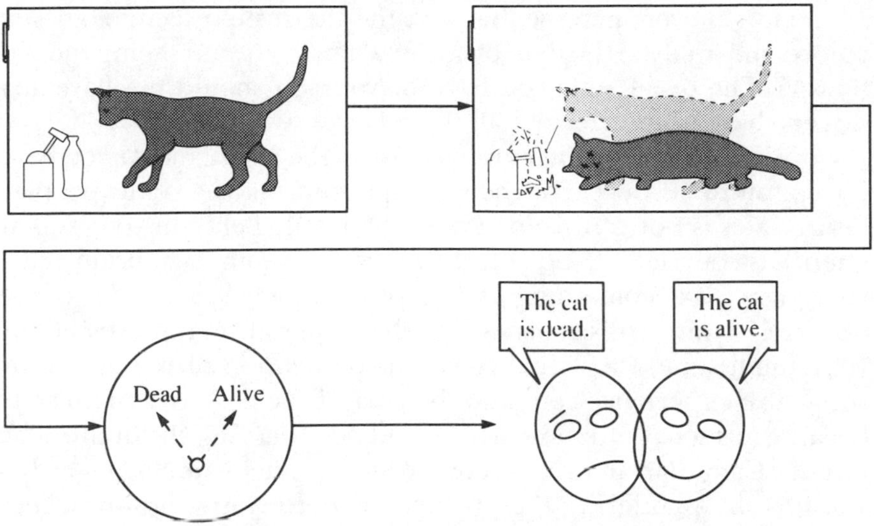

Figure 26. The von Neumann chain. Following von Neumann's argument, even our brain-mind catches the dichotomy of the cat, so how does the chain terminate? (Reprinted from A. Goswami, *Quantum Mechanics*; permission granted by Wm. C. Brown, Inc., publisher.)

Ramachandran's Paradox

If it still bothers you that consciousness is transcendent, you may enjoy examining a paradox that was constructed by the neurophysiologist V. S. Ramachandran.[15]

Suppose that with some supertechnology it is possible to record with microelectrodes, or some such thing, everything that happens in the brain when bombarded by an external stimulus. From such data plus some supermathematics, you can imagine coming up with a complete and detailed state description of the brain under the given stimulus.

Suppose the stimulus is a red flower and that you show it to several people, collect the data, analyze it, and come up with a set of brain states that corresponds to the perception of a red flower. You would expect that, except for minor statistical fluctuations, you would come up with essentially the same state description (something like, certain brain cells in a certain area of the brain involved in color perception have responded) each time.

You might even imagine that with the aid of supertechnology you record and analyze the data of your own brain (upon seeing the red flower). The brain state you find for yourself should not have any discernible difference from all the others.

Consider the following curious twist to the experiment: You have no reason to suspect that the description of all the other people's brain states is not complete (especially if your belief in your superscience is complete). And yet, with regard to your own brain state, you know that something is left out: namely, your role as the observer—your consciousness of the experience represented by your brain state, the actual conscious perception of redness. Your subjective experience could not be part of the objective brain state because in such a situation, who would be observing the brain? The famous Canadian neurosurgeon Wilder Penfield similarly was bewildered by pondering the prospect of performing brain surgery on himself: "Where is the subject and where is the object if you are operating on your own brain?"[16]

There must be a difference between your brain as the observer and the brains of those whom you observe. The only alternative conclusion is that the brain states that you constructed even with superscience are incomplete. Since your brain state is incomplete and other people's brain states are identical to yours, then they must also be incomplete, for they all leave out consciousness.

This is a paradox for the material realists because, from their viewpoint, neither of the above outcomes is desirable. The materialist will be reluctant to give a special privilege to a particular observer (that would amount to solipsism) yet also averse to admitting that any achievable description of the brain state using materialistic science would be, ipso facto, incomplete.

The paradox is resolved by the idealist interpretation of quantum mechanics because in that interpretation the quantum-mechanical description of the brain-mind does not include the transcendent subject, consciousness, and is admitted to be incomplete to that extent. In that incompleteness, room is made for conscious experience.

An important key is the neurosurgeon's question, Where is the subject and where is the object if you are operating on your own brain? The point is made by the expression "what we are looking for is what is looking." Consciousness involves a paradoxical self-reference, an ability, taken for granted, to refer to ourselves separate from the environment.

Erwin Schrödinger said: "Without being aware of it and without being rigorously systematic about it, we exclude the Subject of Cognizance from the domain of nature that we endeavor to understand."[17] A quantum measurement theory that dares to invoke consciousness in the affairs of quantum objects, in order to be "rigorously systematic," must deal with the paradox of self-reference. Let us elaborate on this concept.

When Is a Measurement Complete? [Reprise]

A subtle criticism can be made of the assertion that a transcendent consciousness collapses the wave function of a quantum object. The criticism is that the consciousness that causes the collapse of the wave function might be that of an external, omnipresent God, as in the following:

There once was a man who said, "God
Must think it exceedingly odd
If he finds that this tree
Continues to be
When there is no one about in the quad."

Dear sir, your astonishment's odd
I am always about in the quad
And that's why the tree
Will continue to be
Since observed by, Yours faithfully, God.[18]

An omnipresent God collapsing the wave function does not resolve the measurement paradox, however, because we can ask, At what point is the measurement complete if God is always looking? The answer is crucial: *The measurement is not complete without the inclusion of the immanent awareness.* The most familiar example of an immanent awareness is, of course, that of a human being's brain-mind.

When is a measurement complete? When the transcendent consciousness collapses the wave function by means of an immanent brain-mind looking on with awareness. This formulation agrees with our commonsense observation that there is never an experience of a material object without a concomitant mental object, such as the thought I see this object, or without, at least, awareness.

Note that we have to make a distinction between consciousness with awareness and without awareness. The collapse of the wave function takes place in the former case but not in the latter. Consciousness without awareness is referred to as unconscious in the psychological literature.

Of course, there is some causal circularity to the view that immanent awareness is needed to complete the measurement, since without the completion of the measurement there can be no immanent awareness. Awareness or measurement, which comes first? Which is the first cause? Are we stuck with a chicken-or-the-egg question?

A Sufi story has a similar flavor. One night the Mulla Nasruddin was traveling a lonely road when he spotted a troop of horsemen approaching. The Mulla became nervous and started to run. The horsemen saw him running and went after him. Now the Mulla became really fearful. Coming on the walls of a graveyard and propelled by fear, he jumped the wall, found an empty coffin, and lay down in it. The horsemen had seen him jump the wall, and they followed him into the graveyard. After a little search they found the Mulla looking fearfully up at them.

"Is there anything wrong?" the horsemen asked the Mulla. "Can we help you in any way? Why are you here?"

"Well, it's a long story," replied the Mulla. "To make it short, I am here because of you, and I can see that you are here because of me."

If we are stuck with only one order of reality, the physical order of things, then there is a genuine paradox here for which there is no solution within material realism. John Wheeler has called the circularity of quantum measurement a "meaning circuit,"[19] which is a very sensitive description, but the real question is, Who reads the meaning? Only for idealism is this no paradox, because consciousness acts from outside the system and completes the meaning circuit.

This solution is similar to that of the so-called prisoner's problem, an elementary problem of game theory.[20] Through a tunnel dug with the help of an outside friend, you plan to escape from a prison cell (fig. 27). Obviously, your escape will be much facilitated if both you and your friend dig from opposite sides of the same corner; communication is not possible, however, and there are six corners from which to choose. The chance of escape does not look good, does it? But consider for a moment the shape of your cell, and the chance is excellent that you will choose to dig at corner number 3. Why? Because number 3 is the only corner that looks different (concave) from the outside. Therefore, you would expect your

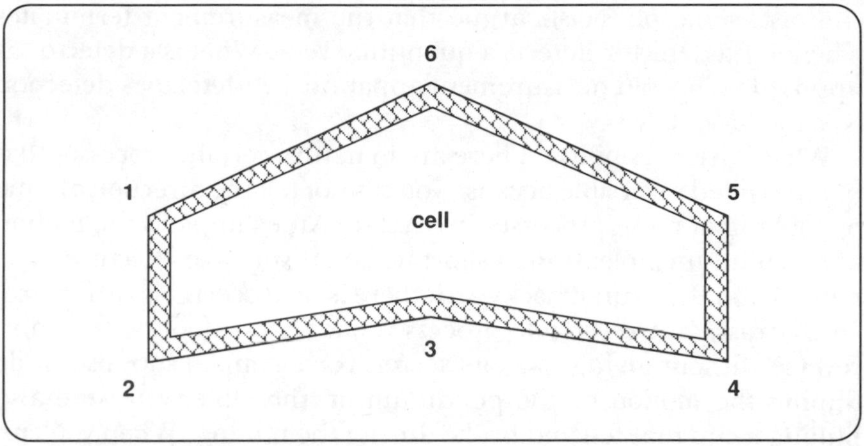

Figure 27. The prisoner's dilemma: Which corner to choose?

friend to begin digging there. Similarly, only number 3 is convex from the inside, so your friend will probably expect you to begin digging there as well.

Now what is your friend's motivation to dig at this particular corner? It is you! He sees you choosing this corner for the same reason that you see him choosing it. Notice that we can assign no causal sequence in this case and therefore no simple hierarchy of levels. Instead of causal linearity, we have causal circularity. No one decided on the plan. Instead, the plan was a mutual creation guided by a higher purpose—the prisoner's escape.

Douglas Hofstadter has called this kind of situation a tangled hierarchy—a hierarchy that is so mixed up that we cannot tell which is higher and which is lower on the hierarchical totem pole. Hofstadter thinks that self-reference may come out of such a tangled hierarchy.[21] I suspect that the situation in the brain-mind, with consciousness collapsing the wave function but only when awareness is present, is a tangled hierarchy and that our immanent self-reference is of tangled hierarchical origin. An observation by a self-referential system is where the von Neumann chain stops.

Irreversibility and Time's Arrow

When is a measurement complete? The idealist says that it is complete only when a self-referential observation has taken place. In contrast, some physicists argue that the measurement terminates whenever a detector detects a quantum event. What is a detector as opposed to any old measurement apparatus? A detector's detection is *irreversible*, they say.

What is irreversibility? There are in nature certain processes that may be called reversible because you cannot tell the direction of time by looking at these processes in reverse. An example is the motion of a pendulum (at least for a short while); if you take a picture of its motion and then run it backward, there is no discernible difference. In contrast, an irreversible process is one that cannot be filmed in reverse without giving away its secret. For example, suppose while filming the motion of the pendulum on the table, you were also filming a cup that fell and broke during the filming. When you run the film in reverse, the fragments of the cup jumping off the floor and becoming whole again will give away your secret—that you are running the film in time-reverse.

To see the difference between a reversible measurement apparatus and a detector, consider an example. Photons have a two-valued characteristic called polarization: an axis that lies along (or is polarized along) only one of two perpendicular directions. Polaroid sunglasses polarize ordinary unpolarized light. They transmit only those photons that have a polarization axis parallel to that of the glasses. To test this, hold two polaroid glasses perpendicular to each other and look through them. You will see only darkness. Why? Because one polaroid lens polarizes the photons vertically (say), but the other lens transmits only photons polarized horizontally. In other words, the two lenses together act as a double filter that screens out all light.

A photon polarized at an angle of 45 degrees to the horizontal is a coherent superposition of half vertically polarized and half horizontally polarized states. If the photon passes through a polarizer-box with both horizontal and vertical polarization channels, it emerges at random either in the vertically polarized or in the horizontally polarized channel. This can be seen from pointer readings on detectors placed behind each channel (fig. 28a).

Now suppose that in the arrangement of figure 28a, we place a 45-degree polarizer in front of the photons before they are detected (fig. 28b). The photon is found to be reconstructed back to its original state of 45-degree polarization, a coherent superposition; it is regenerated. Thus the polarizer alone is not enough to measure the photons—since the photons still retain their potential to become

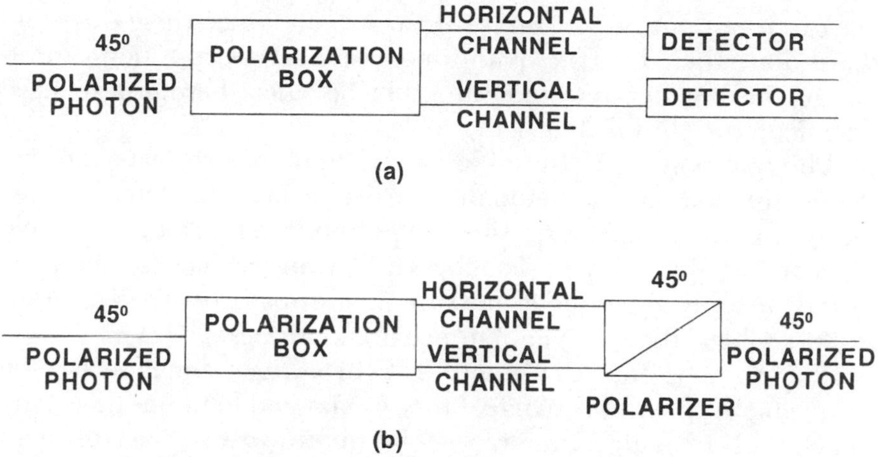

Figure 28. Experiments with 45-degree polarized photons.

a coherent superposition. A detector where irreversible processes take place, such as a fluorescent screen or a photographic film, is needed for measurement.

If you think in terms of time-reversal, the motion of the photons polarized at 45 degrees passing through the polarizer box and then again through the 45-degree polarizer is time reversible. If, however, the photons are detected by some detector with irreversible processes, when you imagine the process backward, you are able to discern between forward and backward.

Recall the story about a scene filmed for a silent movie. The heroine was supposed to be tied to a railroad track while a train sped toward her. In the movie's story line she would be saved—the train would stop just in the nick of time. Since the actress (understandably) was reluctant to risk her life, the director shot the whole scene backward—starting with the actress tied to the tracks while the train was next to her in full stop. Then the train was run backward. But what do you think people saw when the film was run in reverse? In those days trains were fueled by coal-burning boiler. In the backward-running film, the smoke flew into the stack instead of flowing out and thus gave away the secret of the film. The time evolution of smoke is irreversible.

Does this mean that a solution to the problem of quantum measurement is at hand—and without assuming the involvement of consciousness? We have only to recognize the irreversibility of certain measurement apparatuses called detectors, and then perhaps we can jump out of the von Neumann chain. Once these detectors have done their job, the quantum coherent superposition can no longer be regenerated and can truly be said, therefore, to have terminated.[22] But is that really so?

The question is, Is the detector enough to terminate the von Neumann chain? Von Neumann's answer is no. The detector must become a coherent superposition of pointer readings for the simple reason that ultimately, it, too, obeys quantum mechanics. The same is true for any subsequent measurement apparatus—reversible or "irreversible," the von Neumann chain continues.

The point is that the quantum Schrödinger equation is time reversible: It does not change if time is changed to minus time. Any macrobody obeying a time-reversible equation cannot be truly irreversible in its behavior, as shown by the mathematician Jules-Henri Poincaré.[23] Thus the conventional wisdom arises that absolute irreversibility is impossible; the apparent irreversibility that we see in nature has to do with the small probability that exists for a complex macrobody to retrace its path of evolution back to an initial configuration that has more relative order.

Considering irreversibility yields an important lesson. Although ultimately, all objects are quantum objects, the apparent irreversibility of some macro objects enables us to distinguish approximately between classical and quantum. We can say that a quantum object is one that regenerates, while a classical object has a long, long regeneration time. In other words, while quantum objects have no discernible retainment of their history—no memory—classical objects such as detectors can be said to have a memory in the sense of requiring a long time to erase the memory.

Another important issue arises: If there is no ultimate irreversibility in the motion of matter, how does the idealist interpretation handle the notion of unidirectional flow of time, time's arrow? In the idealist interpretation, time is a two-way street in the transcendent domain, showing signs of only approximate irreversibility for motion of more and more complex objects. When consciousness collapses the wave function of the brain-mind, it manifests the subjective one-way time that we observe. Irreversibility and time's arrow enter nature in the process of collapse itself, in quantum measurement, as the physicist Leo Szilard suspected long ago.[24]

It would seem that irreversibility of detectors does not solve the problem of measurement. Such a solution cannot be invoked unless we are ready to accept irreversibility, in the form of randomness, as being even more fundamental than quantum mechanics. There is a proposal to do just that.[25]

Suppose that matter is fundamentally random and that the random behavior of a substratum of particles, through occasional fluctuations, generates the approximate orderly behavior that we may call quantum. If such were the case, quantum mechanics itself would be an epiphenomenon—as would all other orderly behavior. No experimental data support such a theory, although it would be an ingenious solution to the measurement problem if it could be proven. Some physicists do assume, however, that an underlying medium exists that causes the randomness; they draw an analogy with the underlying random motion of molecules that causes the random motion (called Brownian motion) of pollen grains in water when seen under a microscope. The assumption of an underlying medium, however, runs contrary to Aspect's experiment, unless it accommodates nonlocality. It is hard to accept nonlocal Brownian motion within material realism.

The Nine Lives

Stephen Hawking says: "Every time I hear about Schrödinger's cat, I want to reach for my gun." Almost every physicist has had a similar impulse. Everyone wants to kill the cat—the paradox of the cat, that is—but it seems to have nine lives.

In the first life, the cat is treated statistically, as part of an ensemble. The cat is offended (because its singularity is denied in this ensemble interpretation) but not wounded.

In the second life, the cat is viewed as an example of the quantum/classical dichotomy by the divisive philosophers of macrorealism. The cat refuses to trade its life/death dichotomy for another dichotomy.

In the third life, the cat is confronted with irreversibility and randomness, but the cat says, Prove it.

In the fourth life, the cat confronts the hidden variables (the idea that its state never becomes dichotomous but is really completely determined by hidden variables) and what happens is still hidden.

In the fifth life, the neo-Copenhagenists try to do away with the cat using the philosophy of logical positivism. By most judgments, the cat escapes unscathed.

In the sixth life, the cat encounters many worlds. Who knows, it may have perished in some universe, but as far as we can tell, not in this one.

In the seventh life, the cat meets Bohr and his complementarity, but the question What constitutes a measurement? saves it.

In the eighth life, the cat meets consciousness (of a dualistic vintage) face-to-face, but Wigner's friend saves it.

Finally, in the ninth life, the cat finds salvation in the idealist interpretation. Here ends the story of the nine lives of Schrödinger's cat.[26]

Chapter 7

I CHOOSE, THEREFORE I AM

WE HAVE NOT YET CONFRONTED the important question What is consciousness? And how does one distinguish between consciousness and awareness?

Alas, a definition of consciousness is not easy. The word *consciousness* derives from two words: the Latin verb *scire*, which means to know, and the Latin preposition *cum*, which means with. Thus consciousness, etymologically, means "to know with."

In the *Oxford English Dictionary*, moreover, there are not one but six definitions of the word *consciousness*:

1. Joint or mutual knowledge.
2. Internal knowledge or conviction, especially of one's own ignorance, guilt, deficiencies, and so forth.
3. The fact or state of being conscious or aware of anything.
4. The state or faculty of being conscious as a condition or concomitant of all thought, feeling, and volition.
5. The totality of the impressions, thoughts, and feelings which make up a person's conscious being.
6. The state of being conscious regarded as the normal condition of healthy waking life.

None of these definitions is completely satisfactory, but considered all together they provide an approximate understanding of what consciousness is. Imagine a situation in which each of these different definitions comes into play. (We shall assign each definition a subscript—1 through 6.) A bouquet of roses is delivered to you. The delivery man, you, and the sender all share $consciousness_1$ regarding the gift of roses. It is in your $consciousness_2$ that you know the history, associations, and connotations of roses and of their meaning as a gift to you (and in this consciousness, you may or may not appreciate the gift). Your sensory experience of roses resides in your $consciousness_3$, whereby you are able to smell their fragrance, see their color, and feel their thorns. It is your $consciousness_4$, however, that enables you to attach the meanings, consider the relationships, and make the choices connected to the gift (whether to accept or reject the roses, for example). Your $consciousness_5$ is what makes you the unique you, as distinct from your lover and from everyone else, who responds in a particular way to the gift of roses. It is only by virtue of your $consciousness_6$ that you are able to receive the roses, anyway, or to experience or exhibit any of the preceding states of consciousness.

Even this analysis of the word leaves out quite a bit. Consciousness has four different aspects. First, there is the field of consciousness, sometimes referred to as the mind field or global workspace.[1] This is what I have called awareness. Second, there are objects of consciousness, such as thoughts and feelings, that arise and pass away in this field. Third, there is a subject of consciousness, the experiencer and/or witness. (The dictionary definitions are really about the subject of consciousness or the conscious self with which we identify.) Fourth, in idealist philosophy, we speak of consciousness as the ground of all being.

A commonsense definition of consciousness equates it with conscious experience. Speaking of a subject of consciousness without speaking of experience is like speaking of a ballet stage without the ballet. Notice that the concept of conscious experience is not restricted to waking consciousness. Dreaming is a conscious experience, though different from that of the waking state. The states that we experience in meditation, under drugs, in hypnotic trances—all such altered states of consciousness involve experiences.

Common sense also tells us that conscious experiences come with many concomitants, some internal and some external. As I type this page, for example, I watch my mind as my fingers punch the typewriter keys. I am thinking, How well is the page turning out? Should I reword that sentence? Am I explaining too little or too much? And now I hear a knock at my study door. I call out, Who is it? No answer. I have to make a choice. Either I yell louder this time, or I get up and open the door.

Now the external concomitants are easy. I do not identify myself with my fingers, even when they are busy doing things that I value, such as typing this page. Few of us would think of identifying consciousness with sensations, sense impressions, or motor actions. Can you imagine saying, I am my walking to the door? Of course not. Common sense tells us that the external concomitants of a conscious experience are not the fundamental elements of consciousness.

When it comes to the internal stuff of the mind—thoughts, feelings, choices, and so forth—things become much less clear. For example, many people—following the lead of Descartes—identify themselves with their thoughts: I think, therefore I am. For others, being conscious is synonymous with feeling: I feel, therefore I am. Some of us may even identify ourselves with the ability to choose. Nietzsche, for example, equates being and will.

Science is uncommon sense: We resort to science when common sense fails. Turning to psychology does not help, however. As the prominent cognitivist Ulric Neisser said: "Psychology is not ready to tackle the issue of consciousness." Fortunately, physics is. This means returning to quantum theory and to the measurement problem that raised the discussion of consciousness in the first place.

The idealist resolution of the paradox of Schrödinger's cat demands that the consciousness of the observing subject choose one facet from the multifaceted dead-and-alive coherent superposition of the cat and thus seal its fate. The subject is the chooser. It is not *cogito, ergo sum*, as Descartes thought, but *opto, ergo sum*: I choose, therefore I am.

> Mind and mind's laws lay hid in night.
> God said: "Let Descartes be," and there was light.
> It did not last. The devil shouted, "Ho!
> Here's Schrödinger's cat! Restore the status quo."

(Our apologies to Mr. Pope, of course.)

I know, the devotees of classical physics will shake their heads with disapproval because they think that there is no freedom of choice, or free will, in our deterministic world. Because of their assumption of causal determinism, they have attempted to condition us into believing that we are material machines. Suppose that we suspend our conditioning for a few moments. After all, we solved the Schrödinger's cat paradox with our hypothesis.

In the same spirit of investigation, we ask, what then? In answer, a door opens. As captivated as we are with thoughts and feelings, they derive from old, fixed, learned contexts. Is the same true of free will? Our choices set the context for our action, thus the possibility of a new context arises when we choose. It is just this possibility of jumping out of the old context and into a new one at a higher level that makes us free in our choice.

A distinctive language has developed for describing specifically this kind of situation—a hierarchical structure of contextual levels. This language, known as the theory of logical types, was originally developed by Bertrand Russell to solve problems that arose in set theory. Russell's basic idea was that a set consisting of members of the set is of a higher logical type than the members themselves because the set defines the context for thinking about the members. Similarly, the name of a thing, which depicts the context of the thing it describes, is of a higher logical type than the thing itself. Thus, out of the three internal concomitants of conscious experience, choice does stand out. It is of a higher logical type than thoughts and feelings.

Is it the capacity for choice, then, that makes us conscious of the experiences that we choose? In every moment, we literally face myriad alternative possibilities. From these we choose, and as we choose, we recognize the course of our becoming. Thus our choosing and our recognition of choice defines our self. The primary question of self-consciousness is to choose or not to choose.

The idea that choice is the defining concomitant of self-consciousness has some experimental support. Data from experiments in cognitive science indicate that thoughts and feelings, but not choice, arise in response to unconscious perception of stimuli. According to the data, which are described in the following section, we do not seem to exercise choice unless we are acting consciously—with awareness as subjects.

This raises the question of what it means to act without awareness—the concept of the unconscious. What is the unconscious in us? The unconscious is that for which there is consciousness but no awareness. Note that there is no paradox here because in the philosophy of idealism consciousness is the ground of being. It is omnipresent, even when we are in an unconscious state.

Part of the confusion over the term *unconscious perception* arises from the historical idiosyncracies of the term's etymology. It is our conscious self that is unconscious of some things most of the time and of everything when we are in dreamless sleep. In contrast, the unconscious seems to be conscious of all things all of the time. It never sleeps. That is to say, it is our conscious self that is unconscious of our unconscious, and the unconscious that is conscious—we have the two terms backward. Read Daniel Goleman's *Vital Lies, Simple Truths* for further elucidation of this point.

So, when we speak of unconscious perception, we are speaking of events that we perceive but that we are not aware of perceiving.

Unconscious Perception Experiments

I know it sounds odd. How can there be a phenomenon called unconscious perception? Is perception not synonymous with awareness? The writers of the *Oxford English Dictionary* apparently think so. And yet, new data in the cognitive laboratory point toward a distinction between the two concepts—perception and awareness.

The initial experimentation involved two monkeys. Researchers Nick Humphrey and Lewis Weiskrantz had removed the cortical areas connected with vision from the monkeys. Since cortical tissue does not grow back, these monkeys were expected to remain permanently blind. Yet they gradually seemed to recover enough of their sight to convince the researchers that they could see.

One of the monkeys, Helen, was often taken outside on a leash. She gradually learned to do some rather unusual things for a creature who should have been blind. For example, she could climb trees. She also took proffered food when it was near enough to grab but ignored it when it was too far to reach. Clearly, Helen was seeing, but with what?

It turns out that there is a secondary pathway for optical stimuli from the retina to a structure in the hindbrain called the superior colliculus. This collicular vision was enabling Helen to see with what the researchers dubbed blindsight.[2]

By chance, Nick Humphrey came across a human subject with blindsight.[3] A failure in this man's cortex had caused him to become blind in the left visual field of both eyes. Now the experimenters were able to ask the subject what was happening in consciousness when he did certain tasks permitted him by blindsight. And the answers were strange.

For example, if the man was shown a light to his left, his blind side, he could point to it with accuracy. He could also distinguish crosses from circles and horizontal lines from vertical ones, all in his left visual field. But when asked how he saw these things, the fellow insisted that he did not. He claimed that he just guessed, in spite of the fact that his accuracy rate was far beyond that attributable to chance.

What does all this mean? There is now some consensus among cognitive scientists that blindsight is an example of unconscious perception—perception without awareness of it. So you see, perception and awareness are not necessarily intertwined.

Further physiological and cognitive evidence for unconscious perception has come from research done both in America and in Russia.[4] Researchers have measured the electrical responses in the brains of various subjects to a variety of subliminal messages. The responses were usually stronger when a meaningful picture, such as that of a bee, was flashed on a screen for a thousandth of a second than when a more neutral picture, such as an abstract geometrical figure, was shown. (Obviously mathematicians were not in the test group.) Furthermore, when subjects were asked to tell the researchers all the words that came to mind after these subliminal exposures, a meaningful picture yielded words that were clearly related to the image flashed. For example, the picture of the bee elicited such words as *sting* and *honey*. In contrast, a geometrical figure elicited hardly anything related to the object. Clearly, there was perception of the picture of the bee, but there was no conscious awareness of that perception.

These experiments have been hailed in the popular press as experimental proof of Sigmund Freud's concept of the unconscious that startled the scientific world at the turn of the century. What, however, is unconscious in us? The unconscious is that for which there is consciousness (as the ground of being), but no awareness and no subject. So in unconscious perception, we are talking about events that we perceive (that is, events that are taken in as stimuli and processed) but that we are not aware of perceiving. In contrast, conscious perception involves taking in stimuli, processing them, and becoming aware of the perception.

The phenomenon of unconscious perception raises a crucial question. Are any of the three common concomitants of conscious experience (thought, feeling, and choice) absent in unconscious perception? The experiment involving subliminal messages suggests that thought is present, since the subjects thought of the words *sting* and *honey* as a consequence of their unconscious perception of the picture of a bee. Obviously, we go right on thinking even in our unconscious, and unconscious thoughts affect our conscious thoughts.

In regard to feeling, an experiment done with split-brain patients has yielded important evidence. In these subjects, the left and right hemispheres of the brain were surgically disconnected except for the cross-connections in the hindbrain centers that are involved in emotion and feeling. When a picture of a nude male model was projected into the right hemisphere of a female subject during a sequence of geometrical patterns, she showed embarrassment by blushing. When asked why, however, she denied being embarrassed. She had no conscious awareness of these inner feelings and could not explain why she blushed.[5] Thus feeling is also present during unconscious perception, and an unconscious feeling can produce an unexplainable conscious feeling.

Finally, we ask, does choice, too, occur in unconscious perception? To find out, we must send an ambiguous stimulus to the brain-mind so that a choice of responses is available. In a relevant cognitive experiment, the psychologist Tony Marcel used polysemous words, words with more than one meaning. His subjects watched a screen as three words in a series were flashed one at a time at intervals of either 600 milliseconds or 1.5 seconds between flashings.[6] The subjects were asked to push a button when they consciously recognized the last word of the series. The original purpose of the experiment was to use the subject's reaction time as a measure of the relationship between congruence (or lack of it) among the words and the meanings assigned to the words in such series as hand-palm-wrist (congruent), clock-palm-wrist (unbiased), tree-palm-wrist (incongruent), and clock-ball-wrist (unassociated). For example, the bias of the word *hand* followed by the flashing of *palm* might be expected to produce the hand-related meaning of *palm*, which then should improve the reaction time of the subject for recognizing the third word, *wrist* (congruence). If the biasing word were *tree*, then the lexical meaning of *palm* as a tree should be assigned, and the meaning-recognition of the third word, *wrist*, should take a longer reaction time (incongruous). Indeed, this was the result.

When, however, the middle word was masked by a pattern so that the subject saw it unconsciously but not consciously, there was no longer any appreciable difference in reaction time between the congruent and the incongruent cases. This should be surprising, because presumably both meanings of the ambiguous word were available to the person, regardless of the biasing context, yet neither meaning was chosen over the other. Apparently, choice is a concomitant of conscious experience but not of unconscious perception. Our subject-consciousness arises when there is a choice made: *We choose, therefore we are.*

It fits. When we do not choose, we do not own up to our perceptions. Thus the man with blindsight denies seeing anything when he avoids an obstacle. The woman with a split cortex blushes but denies feeling embarrassment.

Perhaps cognitive psychology can help explain consciousness, after all—especially if it can be used to test ideas based on the quantum theory of the subject/self. Both quantum theory and these cognitive experiments show that there is a scientific basis for the emphasis that the Western tradition puts on freedom of choice as central to the human experience.

Notice that if the quantum explanation of Marcel's experiment is correct, then the experiment is indirectly demonstrating the existence of coherent superpositions in our brain-mind. Before choice, the state of the brain-mind is an ambiguous state—like that of Schrödinger's cat. In response to a polysemous word, the brain-mind's state becomes a coherent superposition of two states. Each corresponds to a distinct meaning of *palm*: tree or hand. The collapse consists of the choice between one of these states. (There may be some bias for one meaning because of conditioning. For example, a Californian may have a slight preference for the tree meaning of *palm*. In that case, the probability-weighting of the two possibilities would not be equal but would favor the biased meaning. There would be a nonzero probability for the other meaning, however, and there would still be the question of choice.)

I choose, therefore I am. Remember, also, that in quantum theory, *the subject that chooses is a single, universal subject, not our personal ego "I."* Moreover, as an experiment discussed in the next chapter shows, this choosing consciousness is also nonlocal.

Chapter 8

THE EINSTEIN-PODOLSKY-ROSEN PARADOX

THE IDEALIST SCENARIO of quantum collapse hinges on consciousness being nonlocal. So we need to ask whether there is any experimental proof of nonlocality. We are in luck. In 1982, Alain Aspect and his collaborators at the University of Paris-Sud conducted an experiment that conclusively demonstrated quantum nonlocality.

In the 1930s, Einstein helped create a paradox, now famous as the EPR paradox, to prove the incompleteness of quantum mechanics and to bolster support for realism. Given Einstein's philosophical inclinations, EPR might as well have stood for Einstein for the Preservation of Realism. Ironically, the paradox boomeranged against realism, at least against material realism, and Aspect's experiment is part of this turnabout.

Recall the Heisenberg uncertainty principle—at any given time only one of the two complementary variables, position and momentum, can be measured with absolute certainty. This means that we can never predict the trajectory of a quantum object. With two collaborators, Boris Podolsky and Nathan Rosen (the *P* and *R* of EPR), Einstein constructed a scenario that seems to contradict this unpredictability.[1]

Imagine that two electrons, call them Joe and Moe, interact with each other for a time and then stop interacting. These electrons are, of course, identical twins, because electrons are indistinguishable. Suppose that the distances of Joe and Moe from some origin on a certain axis are x_J and x_M, respectively, as they interact (fig. 29). The electrons are moving and, therefore, they have momentum. We can designate these momenta (along the same axis) by p_J and p_M. Quantum mechanics implies that we cannot measure both p_J and x_J or both p_M and x_M simultaneously by virtue of the uncertainty principle. Quantum mechanics does allow us to measure their distance X from each other ($X = x_J - x_M$) and their total momentum P ($P = p_J + p_M$) simultaneously.

When Joe and Moe interact, said Einstein, Podolsky, and Rosen, they become correlated because even though later they stop interacting, measuring the position of Joe (x_J) enables us to calculate exactly where Moe is—the value of x_M—(since $x_M = x_J - X$, X being the known distance between them). If we measure p_J (Joe's momentum), we can determine p_M (Moe's momentum) because $p_M = P - p_J$, and P is known. Thus by carrying out a suitable measurement on Joe, we can determine either the position or the momentum of Moe. If, however, we are making our measurements on Joe at times when Moe is no longer interacting with Joe, these measurements could not possibly have any effect on Moe. Thus values of Moe's position and momentum must be simultaneously accessible.

A correlated quantum object (Moe) must have simultaneous values of both position and momentum, so concluded EPR. This observation supports realism, because in principle we now could

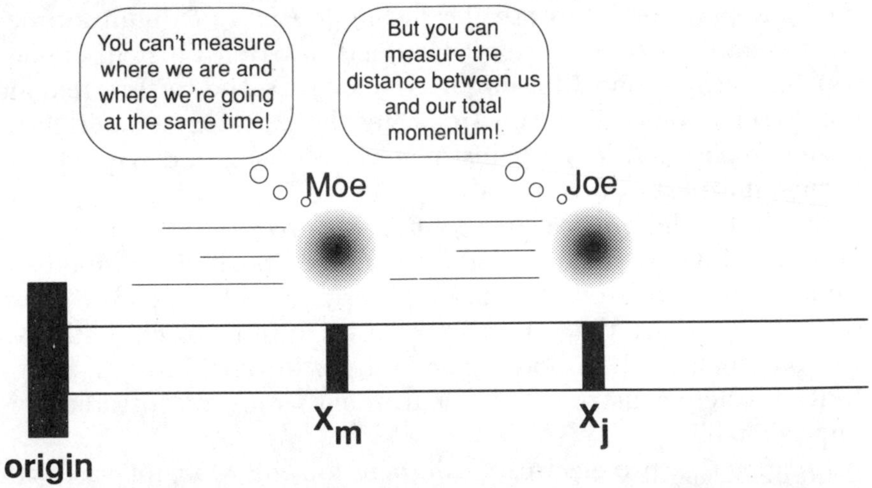

Figure 29. The EPR correlation of Joe and Moe. The distance between them, $x_J - x_M$, is always the same, and their total momentum is always $p_J + p_M$.

determine Moe's trajectory of motion. In contrast, it seems to compromise quantum mechanics severely because quantum mechanics sides with idealism in saying that it is impossible to calculate the trajectory of a quantum object because a trajectory does not exist, only possibilities and observed events exist!

Einstein argued that if the trajectory of a correlated quantum object is, in principle, predictable, but quantum mechanics is unable to predict it, then there must be something wrong with quantum mechanics. Einstein's favorite conclusion from this dilemma was that quantum mechanics is an incomplete theory. Its description of the states of the two correlated electrons is incomplete. Implicitly, he supported the idea that there must be hidden variables behind the scene, unknown parameters that control the electrons and determine their trajectories.

About the concept of hidden variables, physicist Heinz Pagels said: "If we imagine that reality is a deck of cards, all the quantum theory does is predict the probability of various hands dealt. If there were hidden variables, it would be like looking into the deck and predicting the individual cards in each hand."[2]

Einstein supported the idea of deterministic hidden variables in order to demystify quantum mechanics. He was a realist, remember. To Einstein, probabilistic quantum mechanics implied a gambling God, and he believed that God does not play dice. For him it was imperative that quantum mechanics be replaced by some hidden-variables theory in order to restore deterministic order in the world. Unfortunately for Einstein, the difficulty for quantum mechanics that the EPR analysis creates can be resolved without hidden variables, as Bohr first pointed out. Bohr is reported to have said to Einstein: "Don't tell God what to do."

Einstein (and Podolsky and Rosen) were assuming the doctrine of locality in order to revive trajectories and, hence, material realism. Remember, locality is the principle that all interactions are mediated by signals through space-time. Einstein and his colleagues were tacitly assuming the idea that the position (or momentum) measurement on the first electron (the one called Joe) can be done without disturbing the second (Moe), because the two electrons are separated in space and are not interacting via local signals at the time of the measurements. This noninteraction is what we normally expect for material objects, because relativity, with its speed-of-light limit for all signal speeds, prohibits instantaneous interaction at a distance, or nonlocality.

The pertinent issue is separability: Are correlated quantum objects separate when they have no local interaction between them, as objects obeying classical physics certainly are?

Why is the EPR result considered a paradox? The Einsteinian separability is part and parcel of the philosophy of material realism that Einstein defended throughout his later life. This is the philosophy that considers physical objects to be real, independent of each other and of their measurement or observation (the doctrine of strong objectivity). In quantum mechanics, however, the idea of the reality of physical objects independent of our measurement of them is difficult to uphold. Thus Einstein's motive was to discredit quantum mechanics and to re-establish material realism as the underlying philosophy of physics. The EPR paradox says that we have to choose between locality (or separability) and the completeness of quantum mechanics, and this is no choice at all since separability is imperative.

But is it? The answer is a resounding no, for in fact the resolution of the EPR paradox lies in the recognition of an essential inseparability of quantum objects. Measurement of one of two correlated objects affects its correlated partner. This was essentially Bohr's answer to Einstein, Podolsky, and Rosen. When one object (Joe) of a correlated pair is collapsed in a state of momentum p_J, the other's wave function is collapsed also (in the state of momentum $P - p_J$), and we cannot say anything about Moe's position. And when Joe is collapsed by position measurement at x_J, Moe's wave function also collapses immediately to correspond to the position $x_J - X$, and we can no longer say anything about its momentum. The collapse is nonlocal, just as the correlation is nonlocal. EPR-correlated objects have a nonlocal ontological connection, or inseparability, and exert a signal-less instantaneous influence upon each other—as hard as this might be to believe from the point of view of material realism. Separability is the result of collapse. Only after collapse are there independent objects. Thus the EPR paradox forces us to admit that quantum reality must be a nonlocal reality. In other words, quantum objects must be thought of as objects in potentia that define a nonlocal domain of reality that transcends local space-time and thus lies outside of the jurisdiction of Einsteinian speed limits.

Bohr, although he understood inseparability, was reluctant to be explicit about quantum metaphysics. He was not very specific, for example, about what he meant by a measurement. From a fully idealist point of view, we say that a measurement always means an observation by a conscious observer in the presence of awareness. Thus the lesson of the EPR paradox seems to be that a correlated quantum system has the attribute of a certain unbroken wholeness that includes an observing consciousness. Such a system has an innate wholeness that is nonlocal and transcends space.

Before pursuing this line of thinking, we must acknowledge that purely from an experimental point of view, it is difficult to substantiate the correlation of two electrons in the way that the resolution of the EPR paradox demands. Does the wave function of Moe really collapse when we observe Joe at a distance when they are not interacting? David Bohm, a pioneer in deciphering the message of the new physics, thought of a quite practical way to correlate electrons—one that we can use to confirm experimentally the nonlocality of collapse.[3]

An electron has a two-valuedness called spin. Think of spin as an arrow that points up or down on the electron. Bohm suggested that under certain circumstances we can make two electrons collide against each other in such a way that after the collision they would be correlated in that their spin arrows would be pointed opposite to each other. The two electrons are then said to be in a "singlet" state, or correlated in their polarization.

Proving Nonlocality: The Aspect Experiment

Alain Aspect used the singlet kind of correlation between two photons to verify that there is signal-less influence operating between two correlated quantum objects. He verified that a measurement of one photon affects its polarization-correlated partner without any exchange of local signals between them.

Imagine the following setup: A source of atoms emits pairs of photons, and the two photons of each pair move in opposite directions. Each pair of photons is polarization-correlated—their polarization axes lie along the same line. Thus if you see one photon through polaroid sunglasses with the polarization axis vertical (the way they are worn normally), a friend of yours standing at a distance on the opposite side of the light-emitting atoms will see the correlated partner photon only if she likewise is wearing polaroid sunglasses with the axis vertical. If she tilts her head so that her glasses' polarization axis is horizontal, she will not be able to see her photon. If she tilts her head in a way that enables her to see her photon, you will not be able to see its correlated mate because the polarization axis of your glasses is out of sync with the axis of hers.

The photon beams themselves, of course, are unpolarized. They have no particular polarization until you observe them with polaroid glasses; all directions of the beams are equally probable to manifest. Each photon is a coherent superposition of "along" and "perpendicular" polarizations for any direction; it is our observation that collapses a photon with a definite polarization—either along or perpendicular. In a long series of collapses, there will be as many collapses with so-called along polarization as there will be with perpendicular polarization.

Suppose that you both begin with the polarization axes of your glasses vertical so that you each see one of the correlated photons (fig. 30); but then you suddenly tilt your head so that your polarization axis is horizontal instead of vertical. With your maneuver (since

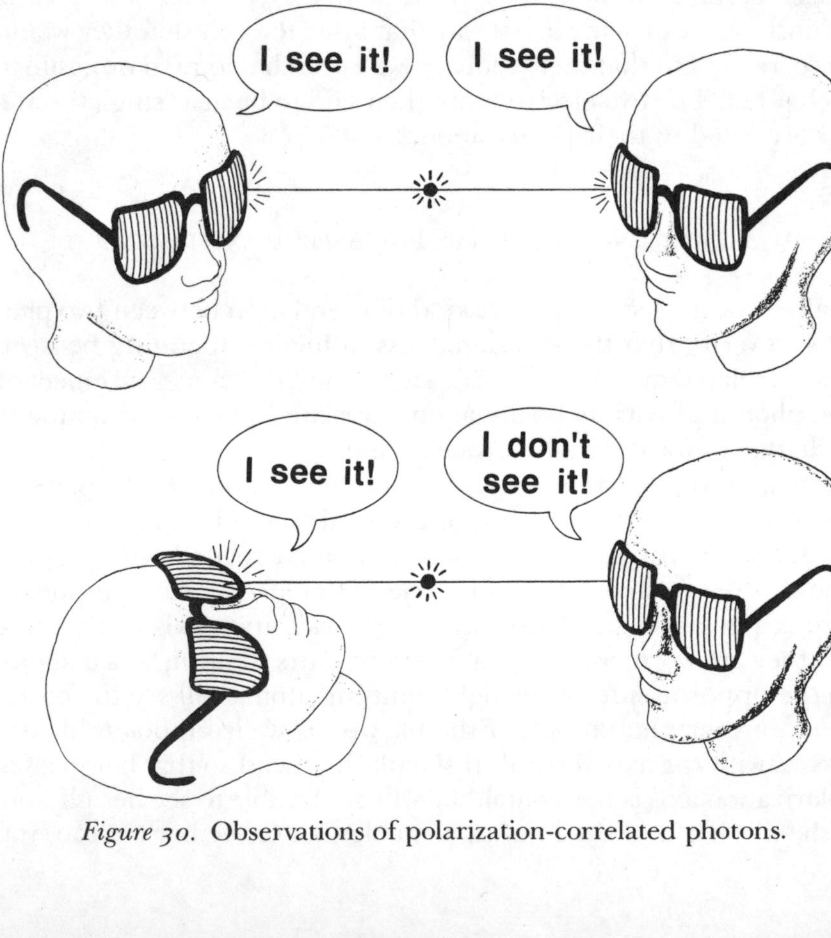

Figure 30. Observations of polarization-correlated photons.

you see the photon only if it is polarized horizontally) you have manifested the photon you see to take on a horizontal polarization axis. Strangely, however, your friend no longer sees the other photon of the pair unless she happens simultaneously to flip her glasses because that correlated photon has also taken on a horizontal polarization axis as a result of your maneuver. This is nonlocal collapse, right?

If you truly believe in material realism, there is something strange about this quantum theoretic construction of events because something you do to one photon is simultaneously affecting its partner at a distance. Whichever direction you flip your sunglasses to see a photon, the correlated partner of that photon always takes on a polarization along the same axis, no matter where it is and no matter how far it is from you. How does the photon know which way to steer unless it is in some sense hearing from its partner? How can it hear spontaneously, defying the speed-of-light limit imposed on signals?

"It is rather discomforting," wrote Erwin Schrödinger in 1935, "that the [quantum] theory should allow a system to be steered or piloted into one or the other type of state at the experimenter's mercy in spite of his having no access to it."[4]

Material realists have worried for the past fifty years about the implication for their philosophy of such strong correlations between quantum objects. Until recently, they could still argue that a local signal between the photons, unbeknownst to us, was mediating the influence, which was thereby strictly obeying realism. Alain Aspect and his collaborators, however, proved in a revolutionary experiment that the influence was instantaneous, occurring without the intermediary of any local signal.[5]

As an example, suppose you are drawing cards from a deck. A friend of yours sitting with his back toward you is telling people what card you are drawing—and he is right every time. This correlation between the two of you might initially be quite disconcerting to onlookers. Eventually, however, people would figure out that somehow you were kicking back a local signal to your friend. That is how many so-called magic tricks work. Now suppose that conditions are arranged in such a way that there is no time for you and your friend to exchange a signal. Still, the correlated magic, your drawing a card and his calling it correctly, keeps happening. This is the peculiar and highly consequential result of Alain Aspect's experiment.

Aspect used polarization-correlated photons that emerged in opposite directions from a source of calcium atoms. A detector was set up on the path of each beam of photons. The crucial feature of the experiment—the one that made its conclusion irrefutable—was the inclusion of a switch that, in effect, changed the polarization setting of one of the detectors every ten-billionth of a second (shorter than the travel time of light or any other local signal between the two detector locations). Even so, the change of the polarization setting of the detector with the switch changed the outcome of the measurement in the other location—just as quantum mechanics says it should.

How did the information about the change in the detector setting pass from one photon to its correlated partner? Certainly not by local signals. There was not enough time for that.

How can one explain it? Consider Pagel's comparison of reality to a deck of cards. The results of Aspect's experiment are like having cards that are being drawn in New York be identical to ones being drawn in Tokyo. We are left with a question: Is the mystery of nonlocality in the cards themselves, or does the consciousness of the observer also come into play?

Material realists reluctantly admit that quantum objects have nonlocal correlations and that if we take the collapse scenario seriously, quantum collapse must be nonlocal. They refuse, however, to see the significance of this and so are missing the most important point of the new physics.

One way to resolve the EPR paradox is by postulating that there is an ether behind the space-time scene where faster-than-light (superluminal) signals are allowed. This resolution would also mean giving up locality and materialism and so is unacceptable to most physicists. Besides, superluminal signals would make possible time travel to the past; this prospect bothers people, and for good reason.

The obvious interpretation of Aspect's experiment is my favorite. According to the idealist interpretation, it is your observation that collapses the wave function of one of the two correlated photons in the experiment, forcing it to take on a certain polarization. The wave function of the correlated partner photon also collapses immediately. A consciousness that can collapse the wave function of a photon at a distance instantly must itself be nonlocal, or transcendent. Thus instead of nonlocality being a property mediated by superluminal signals, the idealist posits nonlocality to be an essential aspect of the collapse of the wave function of the correlated system—and so a trait of consciousness.

So Einstein's hunch about the incompleteness of quantum mechanics, a hunch that was the working hypothesis of the EPR paradox, has produced amazing results. The intuition of a genius is often fruitful in unexpected ways having little to do with the details of the person's theory.

I am reminded of a Sufi story. Mulla Nasruddin was once confronted by a gang of rascals who wanted the shoes that he was wearing. Trying to deceive the mulla, one of the rascals said, pointing to a tree: "Mulla, nobody can climb that tree."

"Of course one can. Here I will show you," said the mulla, taking the bait. Initially the mulla was going to leave his shoes on the ground while he climbed the tree, but on second thought, he tied them together and hung them from his belt. Then he started to climb.

The boys were discouraged. "Why are you taking your shoes with you?" shouted one.

"Oh, I don't know, there may be a road up there, and I may need them!" called down the mulla.

The mulla's intuition told him that the rascals might attempt to steal his shoes. Einstein's intuition told him that quantum theory must be incomplete because it failed to explain correlated electrons. What if the mulla found that there were a road at the top of the tree, after all! This is, in effect, what was discovered by Aspect's experimental study of the EPR paradox.

The Bell Tolls for Material Realism

The paradox of Aspect's experiment is nonlocal collapse. Can we avoid nonlocal collapse by assuming that the photon pairs in the experiment are emitted with definite alignment of their polarization axes? Such is not possible in probabilistic quantum mechanics, but can we invoke hidden variables to do the job? If this eliminates nonlocality, can we then save material realism by invoking hidden variables? No, we cannot. The proof is given by Bell's theorem (named after physicist John Bell, who discovered it), which shows that even hidden variables cannot save material realism.[6]

The hidden variables that Einstein hoped would explain the EPR paradox and restore material realism were, of course, designed to be consistent with locality. They were to act in a local fashion as causal agents on quantum objects, their influence traveling through space-time with a finite velocity during a finite time. Locality of the hidden variables is consistent both with the theory of relativity and with the deterministic belief in local cause and local effect, but it is not consistent with experimental data.

John Bell was the first to suggest a set of mathematical relationships to test the locality of the hidden variables; although these were not equations, they were almost as good. They described a type of relation called inequalities (see reference 6). Aspect's experiment, besides proving that no local signals mediated the connection between EPR-correlated photons, also showed that the inequalities posed by Bell do not hold for real physical systems. Aspect's experiment thus negated the locality of the hidden variables. Not coincidentally, quantum mechanics also predicts that inequalities do not hold for quantum systems. Bell's theorem states that hidden variables, in order to be compatible with quantum mechanics (and with experiment, as it turns out), must be nonlocal.

The far-reaching consequences of the work by EPR and Bell deserve attention. First, study of the paradox pointed out by Einstein, Podolsky, and Rosen revealed the nonlocality of quantum correlations and of quantum collapse. Then, Bell showed that we cannot avoid nonlocality by invoking hidden variables because they, too, exhibit nonlocality; so they cannot save material realism.

Consider physicist Nick Herbert's simple, short, and elegant treatment of a Bell inequality.[7]

Two beams of polarization-correlated photons move in opposite directions from a source. The partners of a correlated pair of photons are called Joe and Moe (*J* and *M*). Two experimenters are set up to observe the J-group and the M-group with detectors made up of calcite crystals that serve as their polaroid glasses. Let's call these calcite crystals *J*-detector and *M*-detector (fig. 31a). As in the similar experiment depicted in figure 30, whenever the *J*-detector and the *M*-detector are set up parallel to each other (that is, with parallel polarization axes) at whatever angle to the vertical, each observer sees one of the correlated photons. When one of the detectors is set at 90 degrees to the other, if one observer sees a photon, the other misses its correlated partner. By definition, if an observer sees a photon, the polarization of the photon is along the polarization axis of his detector calcite crystal (such polarization is denoted by *A*), but if an observer does not see a photon, the conclusion is that the photon is polarized perpendicular to the polarization axis of his calcite crystal (such polarization is denoted by *P*). Notice that now with hidden variables, we are allowing the photons to have definite (correlated) polarization axes independent of our observations. This is the crucial point—with hidden variables, photons have predisposed attributes.

Thus a typical synchronized sequence of detection by two distant observers with parallel settings of their detectors will show a perfect hit pattern, such as the following:

Joe: *A P A A P P A P A P A A A P A P P P*
Moe: *A P A A P P A P A P A A A P A P P P*

And with the detectors at right angles, we will see a perfect sequence of misses, such as:

Joe: *P A P A A P A P P A A A P A P P P A*
Moe: *A P A P P A P A A P P P A P A A A P*

Neither of these results is any longer surprising. Since the photons' polarizations are now predisposed, there is no collapse involved. (Note that the individual beams are unpolarized because in a long sequence each observer sees a 50–50 admixture of *A* and *P* photons.)

We can define a quantity *Polarization Correlation*, or *PC*, that depends on the angle between the detectors. Obviously, if the detectors are exactly at the same angle ($PC = 1$), we have a perfect correlation. If they are at right angles ($PC = 0$), we have a perfect anticorrelation.

At this point Bell asked, What is the value of *PC* for an intermediate angle? Obviously, it has to be between zero and one. Suppose for an angle *A*, the value of *PC* is 3/4. This means that with this setup of the detectors (fig. 31b), for every four photon pairs, the number of hits (on the average) is 3 and the number of misses is 1, as in the following detection sequence:

Joe: *A P P P P A P P A P A A P A A A*
Moe: *A P A P P A A P A P P A P A P A*

If you think of the polarizations as binary-code messages, the messages are no longer the same for the two observers: There is an error (a miss) in Moe's message (compared to Joe's) once in every four observations.

An instance of the inequality relationship described by Bell now becomes apparent. Start with both detectors parallel; the sequences observed are now identical. Change Moe's setting by the angle *A*

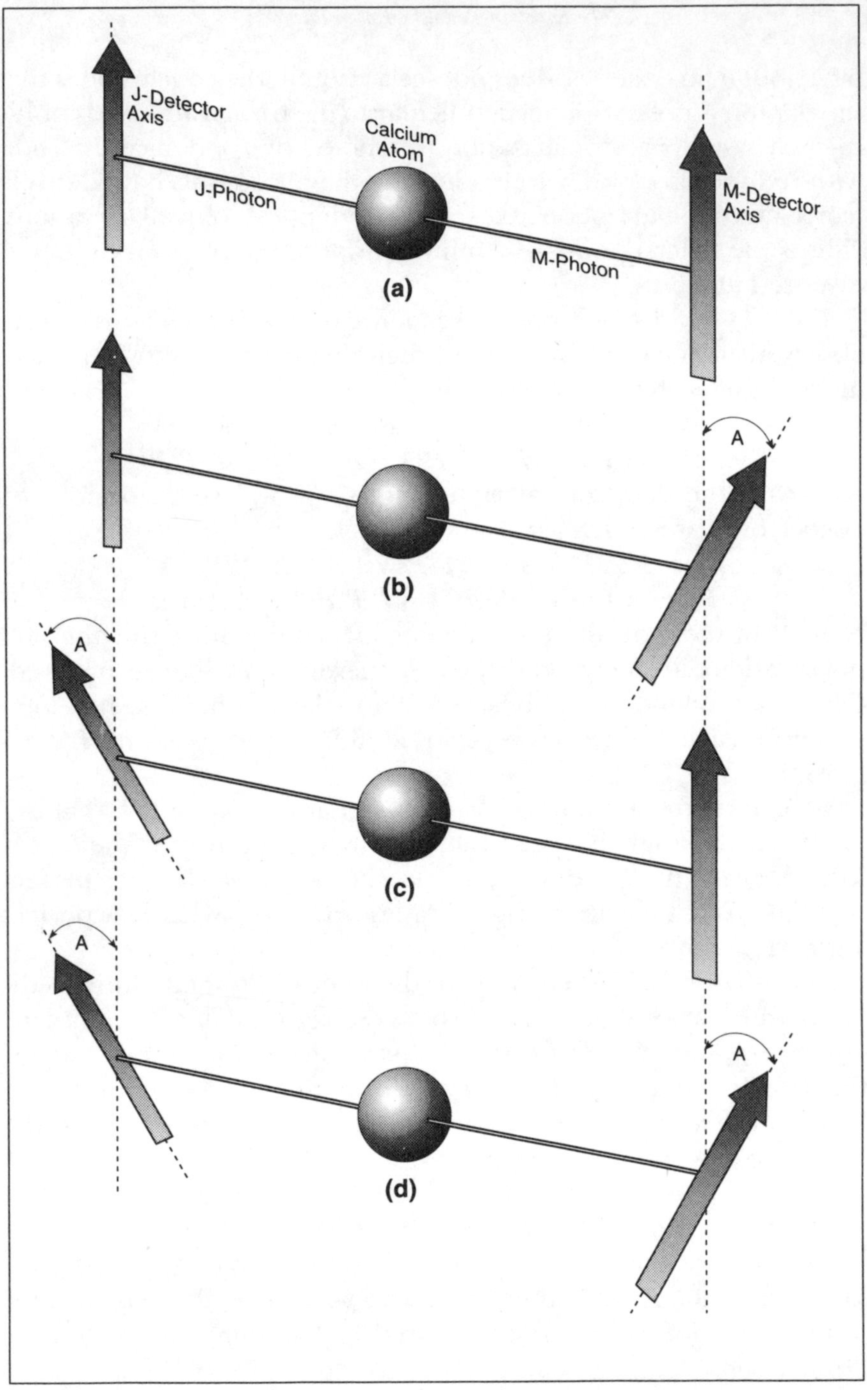
J-Detector Axis
Calcium Atom
J-Photon
M-Detector Axis
M-Photon
(a)
A
(b)
A
(c)
A
A
(d)

(fig. 31b) and no longer are the sequences the same; now they contain errors—one miss in every four observations on the average. Likewise, come back to the parallel setting and this time change Joe's setting by the same angle *A* (fig. 31c); again there will be a miss for every four observations on the average. This result is irrespective of how far apart the two detectors and their observers are. One could be in New York, the other in Los Angeles, with the source somewhere in between.

If locality is valid, if the posited hidden variables that manipulate the photons to take on the particular polarization axis that is demanded by the situation are local, we can say this much with certainty: What you do with Joe's detector cannot alter Moe's message, at least not instantly. And vice versa. Thus, after starting with parallel settings, if the Joe observer turns the Joe detector by the angle *A*, and if at the same time the Moe observer turns the Moe detector in the opposite direction by the same angle (so that the two detectors are now at an angle of 2*A*, fig. 31d), what will the error rate be? If locality of the hidden variables is valid, each maneuver causes an error rate of one out of every four observations, so the total error rate will be two out of four. However, it may happen that every now and then Joe's error cancels out Moe's. Thus the error rate will be less than or equal to 2/4: a Bell inequality. Quantum mechanics, however, predicts an error rate of 3/4 (the proof of which is beyond the scope of this book). This is Bell's theorem: A theory of local hidden variables is not compatible with quantum mechanics.

The Bell inequalities have been investigated experimentally. In 1972, Berkeley physicists John Clauser and Stuart Freedman found that Bell inequalities are indeed violated and that quantum mechanics is vindicated.[8] Then Aspect proved with his experiment that there could be no local signals at all between the two detectors.

Notice how Bell's work (and Bohm's, too, since his work led to the idea of measuring polarization correlation) paved the way to Aspect's experiment that established nonlocality in quantum mechanics. Now you will appreciate why, at a physics conference in 1985, a group of physicists sang to the tune of "Jingle Bells" the following jingle:

Figure 31. How a Bell inequality arises. If the hidden variables are local, the error rate (the deviation from perfect correlation) in arrangement (d) should at most be the sum of the error rates of the two arrangements shown in (b) and (c).

Singlet Bohm, singlet Bell
Singlet all the way.
Oh, what fun it is to count
Correlations every day.

According to Bell's theorem and Aspect's experiment, if hidden variables exist, they must be able to affect correlated quantum objects instantly, even if the objects are separated by an entire galaxy. In Aspect's experiment, when one experimenter changes his detector setting, hidden variables manipulate not only the photon reaching this detector but also its distant partner. Hidden variables are able to act nonlocally. Bell's theorem devastates the local-cause, local-effect dogma of classical physics. Even if you introduce hidden variables to find a causal interpretation of quantum mechanics, as David Bohm does, those hidden variables have to be nonlocal.

David Bohm compares Aspect's experiment to a fish being seen as two distinct pictures in two individual television sets. Whatever one fish does, the other fish does as well. If the fish images are assumed to be the primary reality, this seems strange, but in terms of the "real" fish, it is all very simple.

Bohm's analogy is similar to Plato's allegory of images in the cave, but there is a difference. In Bohm's theory, the light that projects the image of the real fish is not the light of creative consciousness but that of cold, causal, hidden variables. According to Bohm, what happens in space-time is nevertheless determined by what happens in a nonlocal reality beyond space-time. If this were the case, then our free will and creativity would ultimately be illusions, and there would be no real meaning in the human drama.[9] The idealist interpretation promises just the opposite: life is saturated with meaning.

It's a bit like the difference between a movie and a stage improvisation. The action and the dialog in the movie are fixed and determined, but in the live improvisation, variations are possible.

According to the idealist interpretation, violation of the inequalities described by Bell signifies nonlocal correlation between the photons. Hidden variables are not needed as an explanation. Of course, to collapse the wave function of nonlocally correlated photons, consciousness must act nonlocally.

Returning to Bohm's analogy of the fish and its images on two television sets, the idealist interpretation agrees with Bohm in that the fish exists in a different order of reality; that order, however, is a transcendent order in consciousness. The "real" fish is a possibility form already in consciousness. In an act of observation, the fish images simultaneously arise in the world of manifestation as the subjective experience of observation.

Consider another facet of Aspect's experiment. This experiment and the concept of quantum nonlocality have allowed some people to hope that somehow a violation of causality—the idea that cause always precedes effect—is involved. Not necessarily. Since each observer in Aspect's experiment always sees a random 50–50 mixture of *A*s and *P*s, one could never send a message through them. The correlation that we see between the two observers' data appears after we compare the two sets. Only then does its meaning arise in our minds. Thus what Bell's theorem and Aspect's experiment imply is not a violation of causality but that simultaneously occurring events in our space-time world can be related meaningfully to a common cause that resides in a nonlocal realm outside space and time. This common cause is the act of nonlocal collapse by consciousness. (The pattern of meaning being found after the fact is important and will surface again in this book.)

Thus it is not a message transfer that Aspect's experiment indicates but a communication in consciousness, a sharing inspired by a common cause. The psychologist Carl Jung coined the word *synchronicity* to describe meaningful coincidences that people sometimes experience, coincidences that occur without a cause except perhaps a common cause in the transcendent domain. The nonlocality of Aspect's experiment fits Jung's description of synchronicity perfectly: "Synchronistic phenomena prove the simultaneous occurrence of meaningful equivalences in heterogeneous, causally unrelated processes; in other words, they prove that a content perceived by an observer can, at the same time, be represented by an outside event, without any causal connection. From this it follows either that the psyche cannot be localized in time, or that space is relative to the psyche."[10] Jung went on to say, in an insight that we may find startling: "Since psyche and matter are contained in one and the same world and moreover are in continuous contact with one another and ultimately rest on irrepresentable, transcendent factors, it is not only possible but fairly probable, even, that psyche and matter are two different aspects of one and the same thing."[11] This characterization will be useful in our treatment of the brain-mind problem.

If synchronicity still seems like a vague concept, maybe a story will help. A rabbi was walking through the town square when suddenly a man fell on him from a balcony. Because the man's fall was broken by the rabbi, nothing happened to the man; but the poor rabbi's neck was broken. Since this rabbi was a respected wise man who always learned and taught from his own life experiences, his followers asked: "Rabbi, what lesson is there in breaking your neck?" The rabbi answered: "Well, you usually hear, as you sow, so shall you reap. But look what happened to me. A man falls from the balcony, and I break my neck. Somebody sows, and somebody else reaps." This is synchronicity.

It is the same with two correlated photons or electrons or with any other quantum system. Observe one of them and the other is instantly affected because nonlocal consciousness is synchronistically collapsing them both.

Jung had a term for the transcendent domain of consciousness wherein lies the common cause of synchronous events—the collective unconscious. It is called *unconscious* because normally we are unaware of the nonlocal nature of these events. Jung discovered empirically that in addition to the Freudian personal unconscious, there is a transpersonal collective aspect of our unconscious that must operate outside space-time, that must be nonlocal since it seems to be independent of geographical origin, culture, or time.

The nonlocal correlations of Bell's theorem and Aspect's experiment are acausal coincidences, and their meaning—like the events of synchronicity—follows the pattern of always emerging after the fact, when the observers compare their data. If these correlations are examples of Jungian synchronicity, then the aspect of nonlocal consciousness involved here must be related to Jung's concept of the collective unconscious. Our nonlocal consciousness collapses the wave of a quantum object and chooses the result of the collapse when we observe it, but we are normally unaware of the nonlocality of the collapse and of the choice. For further discussion, see chapter 14.

PHYSICS BECOMES A LINK TO PSYCHOLOGY

My interpretation of quantum mechanics is paving the way for the application of physics to psychology. Further debate of this interpretation may be useful, however, since the friction of debate creates illumination.

If we are unaware of the actions of nonlocal consciousness, then is the nonlocal consciousness not perhaps another unnecessary assumption, like that of the hidden variables? Although you can certainly regard nonlocal consciousness as being similar to the hidden variables, you could as easily grant that the idealist's interpretation is suggesting a new way of looking at the hidden variables. Nonlocal consciousness does not constitute causal parameters, as Bohm envisions them, but operates through us; or more properly, it is us—only subtly veiled (a veil that can be penetrated to varied extents, as mystics through the ages testify). Moreover, nonlocal consciousness operates not with causal continuity but with creative discontinuity—from moment to moment, from event to event, as when the quantum wave function of the brain-mind is collapsed. The discontinuity, the quantum jump, is the essential component of creativity; it is precisely the jump out of the system that is needed for consciousness to see itself, as in self-reference.

At one time, probabilistic quantum mechanics encouraged philosophers to look anew at the problem of free will. If you still believe in materialism, however, probability provides only a pale version of free will. When you are stranded at a T-intersection, which way should you go? Is your free choice determined by quantum-mechanical probabilities, or is it the result of some classical determinism playing in your unconscious? The difference is just not that significant. There are other situations where real freedom of choice enters.

Consider creative work. In creativity, we constantly take leaps that catapult us out of the context of our past experiences. In these instances we must exercise the freedom to be open to a new context.

Or consider a case where you have to make a moral decision. Religious creed may suggest that moral values should be dictated by authority, yet looking closely at the process by which human beings make moral decisions, we find that a truly moral decision based on faith and values requires real freedom of choice—a freedom to change the context of the situation.

As an example, consider the struggle for independence from so-called benevolent imperial government. Conventional violent uprisings against rulers rapidly become unethical, don't they? Gandhi succeeded, nevertheless, in ousting the British because he was able to change the context of India's battle for independence by repeatedly using his one weapon: creative choice. His methods were nonviolent protests against the imperialists and noncooperation with the government—methods that were ethical and yet effective.

Most importantly, consider the perception of meaning, which is a common feature of many interesting phenomena in the subjective realm. A book lies on a table in front of you. A person picks it up and utters a meaningless sound, pointedly attracting your attention to the book. Suddenly you understand the meaning of his behavior. He is telling you the word in his language for *book*. How did the meaning of his action arise in your consciousness? It involves nonlocality—a jump out of your local space-time system.

The extraordinary nature of this communication may not be obvious to you because it is so familiar. Imagine, however, that you are the young Helen Keller, deaf and blind since infancy. When Annie Sullivan alternately held Helen's hand under the water and signed the word *water* into her palm, Miss Sullivan was using the same context of communication as in the example involving the word for *book*. Helen must have thought that her teacher was crazy until the meaning of her teacher's actions broke through—until Helen made a jump out of her existing contexts and into a new context.

"The more the universe seems comprehensible, the more it seems meaningless," said the Nobel laureate physicist Steven Weinberg at the conclusion of his popular book on cosmology.[12] We agree. Concepts such as nonlocal and unitive consciousness and the idea of nonlocal collapse make the universe less comprehensible to the materialist scientist. These concepts also make the universe a lot more meaningful to everyone else.

DISTANT VIEWING AS A NONLOCAL QUANTUM EVENT

In the idealist interpretation, the observation of quantum nonlocal correlations is also an unmistakable expression of the nonlocality of consciousness. Can we, therefore, find corroboration of quantum nonlocality in subjective experiences? Is there any such evidence? Yes. The evidence is controversial but interesting.

Suppose that the image of a statue that you have never seen before appears in your mind's eye with such vividness that you can draw a picture of it. Suppose further that a friend of yours is actually looking at the statue at the precise moment that the image appears in your head. This would be telepathy, or distant viewing, and could very well be an example of communication via nonlocal consciousness.

A skeptical scientist might suspect that you knew beforehand what your friend would be viewing. Thus, suppose a couple of researchers ensured with a computer that neither you nor your friend (nor the researchers, for that matter) would know in advance what object would be viewed but only the time at which the telepathic transmission would take place.

The skeptic can still object that a drawing is subject to interpretation. Can you objectively decide whether your drawing indeed represents what your friend saw? So the researchers get disinterested judges—or even better, a computer—to match dozens of your drawings with dozens of distantly viewed sites. The correlation still holds. Would you expect the skeptical scientist to change his view on the subject of telepathy?

Such experiments have been carried out in many different laboratories, and positive results are claimed with both psychic and nonpsychic subjects.[13] Then why has telepathy not yet been recognized as a scientifically plausible discovery? One reason from the scientific point of view is that the data on extrasensory perception (ESP) are not strictly replicable—only statistically so. There is a related apprehension that if ESP were possible, we would be able somehow to transfer meaningful messages through it, a prospect that would create havoc in the orderly world of causality. The most important reason for the skepticism about ESP, however, may be that it does not seem to involve any local signals to our sense organs and hence is forbidden by material realism.

We can attempt to explain the data on distant viewing as experiences of nonlocal correlation that arise in our experience because our mind is quantum. (If you need to do so, suspend your disbelief momentarily.) In terms of Aspect's quantum nonlocality experiment, the question of ESP seems to be one of selection. Only the two correlated psychics, like the two photons in Aspect's experiment, are nonlocally sharing the information. In that experiment, the selection of the experimental design, the source of the photons, and the meaning ascribed to the data reveal that the photons are correlated.[14] Similarly, the correlation of the psychics in a distant-viewing experiment must be related to the preparation of the experiment, the setup, and the meaning ascribed to the data.

Both acausality and meaning in distant viewing (and perhaps in ESP in general) strongly argue in favor of seeing these phenomena as events of synchronicity caused by quantum nonlocal collapse. We cannot custom order synchronicity or acausal phenomena. Remember, the reason that the quantum nonlocal collapse does not conflict with the causality principle is that it precludes dictation of messages.

And so it could be with distant viewing. Perhaps the nonlocal communication between psychics involves no transfer of instrumental information. The correlation between the distant seeing by one psychic and the drawing by the correlated psychic is statistical, and the meaning of the communication becomes apparent only after the drawing is compared to the site viewed. Similarly, in Aspect's experiment, the meaning of the communication between the correlated photons becomes apparent only after the two sets of distant observations are compared.[15]

A recent experiment by the Mexican neurophysiologist Jacobo Grinberg-Zylberbaum and his collaborators directly supports the idea of nonlocality in human brain-minds—this experiment is the brain equivalent of Aspect's (photon) experiment.[16] Two subjects are instructed to interact for a period of thirty or forty minutes until they start feeling a "direct communication." They then enter separate Faraday cages (metallic enclosures that block all electromagnetic signals). Unbeknownst to his or her partner, one of the subjects is now shown a flickering light signal that produces an evoked potential (an electrophysiological response produced by a sensory stimulus and measured by an EEG) in the light-stimulated brain. But amazingly, as long as the partners in the experiment maintain their direct communication, the unstimulated brain also shows an electrophysiological activity, called a transfer potential, quite similar in shape and strength to the evoked potential of the stimulated brain. (In contrast, control subjects do not show any transfer potential.) The straightforward explanation is quantum nonlocality: The two brain-minds act as a nonlocally correlated system—the correlation established and maintained through nonlocal consciousness—by virtue of the quantum nature of the brains.

It is important to note that none of the subjects in the experiment ever reported any conscious experience related to the appearance of the transfer potential. Thus no information at the subjective level was transferred and no violation of the causality principle is involved. The nonlocal collapse and the subsequent similarity of the evoked and transferred potentials of the subjects must be seen as an act of synchronicity; the significance of the correlation is clear only after we compare the potentials. This is similar to the situation in Aspect's experiment.[17]

Can we also find evidence for nonlocality in time? Is there any truth to the so-called precognitive incidents that sometimes become public? For example, there is a claim that somebody foresaw Robert Kennedy's assassination. It is hard to plan a precognitive experiment. Thus I do not see much point in arguing whether a certain psychic did have a genuine precognition or not. There is, however, a clever analysis of the Schrödinger's cat paradox that, at least naively speaking, necessitates the idea of nonlocality in time. According to what we have previously said about consciousness being necessary to collapse the live/dead dichotomy of the cat, the cat remains in limbo until we observe it. Suppose we put lampblack on the floor outside the cage and arrange to have an automatic device open the cage after the hour is up. Suppose we arrive on the scene after another hour passes and find the cat alive. Question: Will the cat's footprints show in the lampblack? If they do, how did the cat make those footprints? An hour ago, the cat was still in limbo. The idea of nonlocality in time provides the easiest way to resolve such a paradox, in the manner suggested by the delayed-choice experiment.

Out-of-Body Experiences

Are there parapsychological phenomena other than distant viewing that may be explained with the quantum/idealist model of consciousness? While it is premature to say definitively that such is the case, there are indications that suggest that we are better off keeping an open mind on the question.

Many people claim that they actually experience going out of their bodies. During such forays they can eavesdrop on friends, watch surgery being performed on themselves, or even travel to distant places.[18] This phenomenon is called out-of-body experience (OBE). The similarity of the OBE to a transmigration of the mind's "I" out of the body is undeniable, but how could that be? It sounds a lot like mind-body dualism.

The validity of the out-of-body experience as a genuine phenomenon of consciousness has gained credibility. Read, for example, Michael Sabom's book *Recollections of Death*, which reports a significant and systematic study of OBE in connection with near-death experiences. As a cardiologist with access to medical charts, Sabom had the unique advantage of being able to verify many technical details in the subject-patients' OBE reports of emergency medical-intervention procedures performed on their virtually dead bodies. His subjects described with great accuracy procedures that were clearly outside their physical bodies' fields of view.

Since these subjects had extensive medical histories involving repeated hospital admissions and experiences with hospital procedures, it would not be too surprising if they were making successful educated guesses. To rule out this possibility, Sabom used a control group of patients with similar medical histories, including near-death crises, but who did not experience OBEs. When asked to describe what they thought happened in the emergency room while they were in their near-death conditions, these control patients gave reports that contained many inaccuracies and very scant correlations, even in a general way, with the facts. Originally skeptical himself, Sabom took great care to conduct his investigations and to evaluate his findings in accord with the rigorous standards of today's psychological laboratory methodology.

Can the mind actually leave the body? In such psychic experiences as OBE it certainly seems that way. This legitimate question cannot be dismissed cavalierly by invoking hallucination, as locality-bound materialist scientists sometimes attempt to do. Sabom, who has quite thoroughly researched the question of whether OBE is hallucination, has this to say: "Unlike the NDE [near-death experience], autoscopic [self-visualizing] hallucinations (1) consist of the physical body ('original') perceiving the projected image ('double'); (2) involve direct interaction between the 'original' and the 'double'; (3) are perceived as being unreal; and (4) commonly evoke negative emotions. For these reasons, the autoscopic hallucination does not appear to be a plausible explanation of NDE."[19]

Quite frankly, when I first looked at OBE in the early eighties, I was impressed with this and other research and started looking for some alternative way of viewing the phenomenon that would enable me to explain it within a scientific framework—without resorting to either hallucination or the transmigration of the mind. Somehow, disembodied minds, or astral bodies as they are called in some circles, watching their physical bodies undergo surgery was to me a noncompelling and simplistic explanation of what I could accept only as a subjective perception of an optical illusion.

An example of a familiar optical illusion may make the distinction clear. I have always been fascinated by the moon illusion: the fact that the horizon moon looks so much bigger to the eye in nature than it does in a photograph. Detailed experiments carried out by scientists, as well as my own nonrigorous fooling around with the phenomenon, have convinced me that the illusion is a size illusion.[20] When the moon is at the horizon, the brain is deceived into perceiving it to be at a greater distance than the overhead moon. The brain therefore compensates, making the image look bigger.

The idea continued to haunt me that OBE must be some sort of an illusion, but of what? Meanwhile, I was also surveying the literature on distant viewing. It suddenly occurred to me that OBE must be an illusory construction of distant viewing, which is nonlocal viewing outside one's physical field of view. Objectively, this is what the near-death subjects of Sabom were doing. But why the illusion of being out of the body?

When very young children see or hear something outside their field of sense perception, they have the reverse difficulty that an adult distant-viewer has. The child's difficulty, one of externalizing the universe, arises from the fact that all our awareness of the external world truly occurs inside our heads because the optical and auditory images are formed inside of our brains. Slowly, using their senses of touch and taste extensively, children learn to externalize the world. They develop perceptual discriminations that enable them to recognize distance effects on viewing and hearing.

For an adult, the unfamiliar experience of the distant viewing of an object outside the visual field must produce considerably more cognitive chaos than a child experiences. The adult's ingrained conditioned system of perception says that the object is somewhere else; therefore, one would have to be "there" to see it. As in the moon illusion, the brain is fooled to construe the nonlocal distant viewing as an out-of-body experience. So if a person is watching her own anesthetized surgery, normally an impossible feat, her soul or astral body must be hovering near the ceiling or across the room—since that is the location from which she seems to be perceiving the action.

Once I saw that OBE could well be a phenomenon of distant viewing, a veil lifted. Here at last was an explanation of OBE that could satisfy the skepticism of a scientist. The nonlocality of our consciousness is the key to resolving the paradox.

Incidentally, if you are skeptical about the nonlocality of distant viewing and feel that some sort of local signals may be involved that we have not yet found, you should know that researchers, especially in Russia, have looked for such signals for years without ever finding any.[21] Some of their experiments involve having psychics demonstrate their ESP from inside Faraday cages, but the Faraday cages seem to have no demonstrable effect on ESP ability.

Furthermore, local signals spread out from their source in the space surrounding it; thus the intensity at a point away from the source attenuates with distance. The farther away the point, the less intense is the signal reaching it. In contrast, nonlocal communication exhibits no such attenuation. Since the evidence indicates that there is no distance attenuation of distant viewing, distant viewing must be nonlocal.[22] Thus it is logical to conclude that psychic phenomena, such as distant viewing and out-of-body experiences, are examples of the nonlocal operation of consciousness.

Any attempt to dismiss a phenomenon that is not understood merely by explaining it as hallucination becomes irrelevant when a coherent scientific theory can be applied. Quantum mechanics undergirds such a theory by providing crucial support for the case of the nonlocality of consciousness; it provides an empirical challenge to the dogma of locality as a universal limiting principle.

Perhaps even more surprising, the nonlocal view of consciousness resolves paradoxes not only of extrasensory perception but also of ordinary perception, as we will see in the next chapter.

It is likely, as it becomes clear that Bell's theorem and Aspect's experiment have really tolled the death of material realism, that the scientist's resistance to the acceptance of the validity of distant-viewing experiments and other parapsychological phenomena will give way. At a recent Physical Society meeting, one physicist was overheard to say to another: "Anybody who is not bothered by Bell's theorem has to have rocks in his head."[23] Even more heartening, a poll of physicists at a conference revealed that a full 39 percent of the physicists assembled were indeed bothered by Bell's theorem. Since such a high percentage of physicists are bothered, we might well expect the idealist paradigm of physics to get a fair hearing.

Chapter 9

THE RECONCILIATION OF REALISM AND IDEALISM

MATERIAL REALISM cannot be saved. Two important questions must then be addressed: First, why does the macro universe look so realistic? Second, without some sort of realism, how can we do science? The resolution is that material realism can be incorporated within monistic idealism. Before we look at how this can be done, let us consider why quantum mechanics requires an interpretation at all. Why do we need philosophy to understand it? Why can it not speak for itself? A summary of the reasons follows:

1. The state of a quantum system is determined by the Schrödinger equation, but the solution to the Schrödinger equation, the wave function, is not directly related to anything we observe. Thus the first question of interpretation is what the wave function represents: a single object? a group of similar events? an ensemble of objects? The square of the wave function determines probabilities, but how should we understand the probabilities? This calls for interpretation. We favor a *single object* interpretation, but it is still a question of philosophy.
2. Quantum objects are governed by the Heisenberg uncertainty principle: It is impossible to measure simultaneously and with certainty pairs of conjugate variables, such as position and momentum. Is this purely a question of measurement (the

effect of quantum probes exerting an uncontrollable amount of energy on the object they measure), or does the uncertainty principle arise out of the nature of things? The uncertainty principle arises from the nature of the wave packets that we have to construct in order to derive localized particles from waves. Again, this answer depends on interpretation and philosophy.

3. The paradox of wave-particle duality—that quantum objects have both wave and particle aspects—needs a resolution, which means interpretation and philosophy.
4. What physical reality, if any, could a coherent superposition have? Can we really resolve the paradox of Schrödinger's cat without seriously considering this sort of question? The consideration of this sort of question invariably involves interpretation and metaphysics.
5. Are discontinuity and quantum jumps truly fundamental aspects of the behavior of quantum systems? In particular, we have portrayed the collapse of a wave function or a coherent superposition in a measurement situation as a discontinuous event. But is collapse necessary? Can we find interpretations that avoid collapse and thus avoid discontinuity? Notice that the motivation for searching for such interpretations is to shore up a philosophical position: that of realism.
6. Bohr's correspondence principle affirms that under certain conditions (for example, for very closely spaced energy levels in atoms) quantum mechanical predictions reduce to those of classical mechanics. This guarantees that we can use classical mechanics to make predictions about macro objects in most situations, but does it ensure that measurement apparatuses behave classically when needed? Some physicists (realists, all) think that this is a question of philosophy.
7. Bell's theorem and the Aspect experiment force us to ask how we should interpret the meaning of quantum nonlocality. This has grave repercussions for our philosophy.

Material realism, stymied by quantum mechanics, gets in trouble whenever the question of the nature of quantum reality comes up—be it in connection with the uncertainty principle, with wave-particle duality, or with coherent superpositions. Whenever we ask if there is some other kind of reality beyond the material reality, we are putting material realism on the spot. Similarly, a genuine discontinuity points to a transcendent order of reality and thus a breakdown of material realism.

The paradoxes of quantum measurement (that of Schrödinger's cat, for example) are impossible difficulties for a material realist. A materially real cat with no other order of reality in which to exist must face the problem of coherent superposition squarely. Can a cat really be dead and alive at the same time?

Finally, the Bell-Aspect nonlocality is the ultimate challenge to material realism. There are only two alternatives, and neither is compatible with strict materialist philosophy. Giving up locality in favor of faster-than-light signals in a realm beyond space-time is obviously a jump beyond the material order, as is the acceptance of nonlocal hidden variables. Giving up strong objectivity or accepting any kind of role for conscious observation relegates material realism to the pile of obsolete theories that include the flat earth, ether, and phlogiston (the never-found substance that was proposed as the active agent in the heat and light of combustion).

Can We Reconcile a Many-Worlds Theory with Idealism?

The various models for resolving the paradox of Schrödinger's cat are all flawed except for three—the many-worlds theory, the theory of nonlocal hidden variables, and the present theory based on monistic idealism. From the discussions in the previous chapter you can see sufficient reasons to question a hidden-variables interpretation. Idealism has a clear edge there. Can the idealist interpretation also claim an edge on the many-worlds theory?

The many-worlds theory attempts to resolve the quandaries that are posed by the paradox of Schrödinger's cat by postulating that the universe splits into two branches: one with a dead cat and a lamenting observer, the other with a live cat and a happy observer. Try, however, to use this theory for resolving the paradox of quantum nonlocality. A measurement here of a correlated electron still splits the world of its partner over there at a distance and yet instantly. Thus this interpretation seems to compromise locality and, hence, does not uphold material realism after all.

Even though it does not help support material realism, the many-worlds theory should certainly be considered a viable alternative to the idealist interpretation. But the many-worlds alternative (like the nonlocal hidden-variables theory) abandons many of the revolutionary aspects of the Copenhagen interpretation. In contrast, monistic idealism takes off from where the Copenhagen interpretation becomes fuzzy; it declares explicitly that the quantum waves, or coherent superpositions, are real but exist in a transcendent domain that is beyond and in addition to material reality.

Actually, the many-worlds idea can be incorporated easily into the idealist interpretation. When we examine the many-worlds theory carefully, we find that it employs conscious observation. For example, how does one define when a branching of the universe occurs? If this happens when there is a measurement, then by the definition of measurement it involves the role of the observer.

According to the idealist interpretation, coherent superpositions exist in a transcendent domain as formless archetypes of matter. Suppose that the parallel universes of the many-worlds theory are not material but archetypal in content. Suppose that they are universes of the mind.[1] Then, instead of saying that each observation splits off a branch of the material universe, we can say that *each observation makes a causal pathway in the fabric of possibilities in the transcendent domain of reality.* Once the choice is made, all except one of the pathways are excluded from the world of manifestation.

Behold how this way of reinterpreting the many-worlds formalism gets rid of the costly proliferation of material universes.

One attractive feature of the many-worlds theory is that the existence of many worlds makes it a little more palatable to apply quantum mechanics to the entire cosmos. Since quantum mechanics is a probabilistic theory, physicists feel uncomfortable thinking about a wave function for the entire universe, such as Stephen Hawking has proposed.[2] They wonder whether one can ascribe meaning to such a wave function if there is only one of a kind. The theory of many-worlds, even in the transcendent domain, helps address this problem.

The truly cosmological question can now be answered: How has the cosmos existed for the past fifteen billion years if for the bulk of this time there were no conscious observers to do any collapsing of wave functions? Very simple. The cosmos never appeared in concrete form and never stays fixed in form. Past universes, one after the other, cannot be seen like paintings on canvases from which present events unravel with time, although if you think about it, this unraveling universe is how material realists picture it.

I propose that *the universe exists as formless potentia in myriad possible branches in the transcendent domain and becomes manifest only when observed by conscious beings.* To be sure, there is the same circularity here that gives rise to the self-reference discussed in chapter 6. It is these self-referential observations that plot the universe's causal history, rejecting the myriad parallel alternatives that never find their way to material reality.

This way of interpreting our cosmological history may help explain a puzzling aspect of the evolution of life and mind, namely that there is only a very low probability for the evolution of life from prebiotic matter through beneficial mutations leading to us. Once we recognize that biological mutation (which includes the mutation of prebiotic molecules) is a quantum event, we realize that *the universe bifurcates in every such event in the transcendent domain, becoming many branches, until in one of the branches there is a sentient being that can look with awareness and complete a quantum measurement.* At this point the causal pathway leading to that sentient being collapses into space-time reality. John Wheeler calls this kind of scenario the closure of the meaning circuit by "observer-participancy."[3] Meaning arises in the universe when sentient beings observe it, choosing causal pathways from among the myriad transcendent possibilities.

If this sounds as if we are re-establishing an anthropocentric view of the universe, so be it. The time and context for a strong anthropic principle has come—the idea that "observers are necessary to bring the universe into being."[4] It is time to recognize the archetypal nature of mankind's creation myths (found in the Book of Genesis in the Judeo-Christian tradition, in the Vedas of the Hindu tradition, and in many other religious traditions). The cosmos was created for our sake. Such myths are compatible with quantum physics, not contradictory.

A great deal of misunderstanding arises because we tend to forget what Einstein said to Heisenberg: What we see depends on the theories we use to interpret our observations. (Of course, Immanuel Kant and William Blake had already told us this, but they were ahead of their time.) How we reconstruct the past always depends on the theories we use. For example, consider how people looked at sunrise and sunset before and after the Copernican revolution. Copernicus's heliocentric model shifted attention away from us—we were no longer the center of the universe. But now, the tide is turning. Of course, we are not the geographical center, but that is not the issue. *We are the center of the universe because we are its meaning.*

The idealist interpretation thoroughly recognizes this dynamic aspect of the past—that the interpretation of what we see changes with our conceptual notions, like a myth.[5] Nor do we have to be chauvinistic: We can as easily suppose that the universe that collapsed into physical space-time reality is the one with the possibility of the evolution of the greatest number of intelligent, self-aware beings on billions and billions of planets throughout the expanding universe.

HOW CAN AN IDEALIST COSMOS CREATE THE APPEARANCE OF REALISM?

If reality consists of ideas ultimately manifested by consciousness, how do we explain so much consensus? If it is idealism that wins the philosophical debate and if realism is a false philosophy, how can we do science? David Bohm has said that science cannot be carried out without realism.

There is some truth to Bohm's statement. But I will present convincing logic that the essence of scientific realism can be incorporated under the broad umbrella of idealism.

To treat this issue in full, consider the origin of the realism/idealism dichotomy in the paradox of perception. The artist René Magritte drew a picture of a pipe, but the caption read: *ceci n'est pas une pipe* (this is not a pipe). Then what is it? Suppose that you say: This is a picture of a pipe. That's a good answer, but if you are a true master of tricks, you will say: I see the image caused in my head (brain) by the sense impressions of a picture of a pipe. Exactly. No one ever saw a picture in an art gallery. You always see the picture in your head.

Of course, the picture is not the object. The map is not the territory. Is there even a picture out there? All we know for sure is that there is some sort of a picture in our brains, a truly theoretical image. In any event of perception it is this theoretical, very private image that we actually see. We assume that the objects we see around us are empirical objects of a common reality—quite objective and public, quite subject to empirical scrutiny. Yet, in fact, our knowledge about them is always gathered by subjective and private means.[6]

Thus arises the old philosophical puzzle about which is real: the theoretical image that we actually see but only privately, or the empirical object that we do not seem to see directly but about which we form a consensus?

The inner privacy of the theoretical image would be no problem, and there would be no discernible dichotomy, if there were always a one-to-one correspondence between that image and an empirical object that others could verify immediately. This is not the case; there are optical illusions. There are creative and mystical experiences of subjective images that do not necessarily correspond to anything in the immediate consensus reality. Thus the authenticity of theoretical images is suspect, and this in turn compromises the authenticity of empirical objects as well because we never experience them without the intermediary of a theoretical image. This is the paradox of perception: We cannot seem to trust the authenticity of either our theoretical image or the consensus, public, empirical object. Philosophical "isms" are born out of such paradoxes.

Historically, two schools of philosophy have debated what is really real. The idealist school believes that the theoretical image is more real and that the so-called empirical reality is but ideas of consciousness. In contrast, realists hold that there must be real objects out there—objects about which we form a consensus, objects that are independent of the subject.

In practice each of these views has its uses. Without some form of realism, some presumption that there are empirical objects that are independent of the observer, science is impossible. Agreed. Without the conceptualization and validation of theoretical ideas, however, science is equally impossible.

Hence we need to transcend the paradox. This was done by the philosopher Gottfried Leibniz and subsequently by another philosopher, Bertrand Russell, with a seemingly absurd idea: Both views can be right if we have two heads, with the empirical object inside one but outside the other.[7] An empirical object would be outside what we might call our small head, and thus realism is validated; the object would simultaneously be inside our big Head and thus be a theoretical idea in this big Head, which would satisfy the idealist. By a clever philosophical maneuver, the object has become at once both an empirical object outside of empirical heads and a theoretical image inside an all-encompassing theoretical Head.

You may ask, is this theoretical big Head just theoretical, or does it have any empirical reality? The plot thickens when we realize that this big Head embraces all empirical small heads and is thus itself the object of empirical scrutiny. Suppose we take the idea of this big Head seriously.

When we look closely, we suspect that the big Head does not have to be separate but can be constituted in all the empirical heads (that is, there is no reason to postulate more than one such Head, since it holds all empirical reality within it; we can all be sharing one Head). Suppose that the head, the brain, is part of a consciousness that has two aspects, two different ways of organizing reality: one aspect that is local, quite confined within the empirical brain, and another global consciousness that encompasses the experience of all empirical objects, including the empirical brains.

You will recognize nonlocality in the last statement. The concept of nonlocality has brought respectability to the seemingly absurd suggestions of Leibniz and Russell. If, in addition to the local ways of gathering data, there is a nonlocal organizing principle connected with the brain-mind, nonlocal consciousness, what then? This is tantamount to our having two heads, and the paradox of perception is solved.[8]

How close our considerations of reality now seem to what the writers of the Upanishads intuited millennia ago:

> It is within all this
> It is outside all this.[9]

What is more, now both idealism and realism can be valid. Both are right. For if the brain-mind itself is an object in a nonlocal consciousness that encompasses all reality, then what we call objective empirical reality is within this consciousness. It is a theoretical idea of this consciousness—thus idealism is valid. When, however, this consciousness becomes immanent as a subjective experience in a part of its creation (in the brain-mind that is localized in our head) and looks through its organization of sense perceptions at other locally separated parts of the creation as objects, then the doctrine of realism is useful for studying the regularities of behavior of these objects.

Now to the important question: Why is there so much consensus? The phenomenal world looks overwhelmingly objective for two reasons. First of all, classical bodies have huge masses, which means that their quantum waves spread rather slowly. The small spreading makes the trajectories of the center of the mass of macro objects very predictable (whenever we look, we find the moon where we expect it), thus producing an aura of continuity. Additional continuity is imposed by our own brain-mind's perceptual apparatus.

Second, and even more important, the complexity of macro bodies translates into a very long regeneration time. This allows them to make memories or records, as temporary as they may be in the final reckoning. Because of these records, we are tempted to look at the world in causal terms, employing a concept of one-way time that is independent of consciousness.

Conglomerates of quantum objects that we can call classical are necessary as measuring apparatuses to the extent that we can define their approximate trajectories and speak of their memory. Without these classical objects the measurement of quantum events in space-time would be impossible.

In the nonlocal consciousness, all phenomena, even so-called empirical, classical objects, are objects in consciousness. It is in this sense that idealists say that the world is made of consciousness. Clearly, the idealist view and the quantum view converge if we accept the nonlocal solution of the paradox of perception.

I trust my intuition that the idealist interpretation of quantum mechanics is the correct one. Of all the interpretations, this is the only one that promises to take physics into a new arena: the arena of the brain-mind-consciousness problem. If history is any guide, all new breakthroughs in physics extend its arena. Can quantum mechanics and the philosophy of idealism together form the basis of an idealist science that can solve the knotty paradoxes of the mind-body problem that have puzzled us for millennia? Yes, I believe they can. In the following part of this book I attempt to set the groundwork for that solution.

Abraham Maslow wrote: "If there is any primary rule of science, it is, in my opinion, acceptance of the obligation to acknowledge and describe all of reality, all that exists, everything that is the case. . . . At its best it [science] is completely open and excludes nothing. It has no 'entrance requirements.' "[10]

With idealist science, we have arrived at a science that has no entrance requirement, that excludes neither the subjective nor the objective, neither spirit nor matter, and thus is able to integrate the deep dichotomies of our thought.

PART 3

SELF-REFERENCE: HOW THE ONE BECOMES MANY

Centuries ago, Descartes portrayed mind and body as separate realities. That dualistic schism still pervades our view of ourselves. In this part, we shall show that a monism based on the primacy of matter is incapable of exorcising the demon of dualism. What does bridge the schism is idealist science—an application of quantum physics as interpreted in accordance with the philosophy of monistic idealism.

We shall see that idealist science not only heals the schism of the mind-body relationship but also answers some questions that have puzzled idealist philosophers for ages—questions like, How does the one consciousness become many? and How does the world of subjects and objects arise from one undivided being? The answers to such questions are found within such concepts as tangled hierarchy and self-reference—a system's capacity to see itself separate from the world.

In India there is a wonderful legend about the origin of the river Ganges. In actuality, the Ganges is born from a glacier high in the Himalayas, but the legend says that the river originates in heaven and comes to earth via the tangled braids of Shiva's hair. A famous Indian scientist, Jagadish Bose, who had far-reaching ideas about the consciousness of plants, wrote in his reminiscences that in his childhood he would listen to the sound of the Ganges and wonder about the meaning of the legend. When he became an adult, he found an answer: cyclicity. Water evaporates and forms clouds, then comes down as snow in the highest peaks of the mountain. The snow melts and becomes the source of rivers, which then find their ways to the ocean, only to evaporate again as the cycle continues.

I, too, as a youth spent hours on the banks of the Ganges pondering the meaning of the legend. Somehow, I did not think that Bose had the final answer to the meaning. Cyclicity, of course, but what was the meaning of Shiva's tangled braid? I did not know the answer, not then.

After looking at many different rivers, the legend continued to mystify me until I read Doug Hofstadter's book Gödel, Escher, Bach: An Eternal Golden Braid. *In the legend, the river Ganges (another name of the divine mother) symbolizes the formless principle behind manifest form, the Platonic archetypes; and Shiva is the formless principle behind manifest self-consciousness, the unconscious. Shiva's tangled braid represents a tangled hierarchy (Hofstadter's eternal golden braid). Reality comes to us in a manifest form via a tangled hierarchy just as the Ganges descends to the world of form via Shiva's tangled braids.*

We will find that this answer leads us to the idea of a spectrum of self-consciousness. We find that there is self beyond ego. Consideration of this greater self enables us to integrate the various personality theories of modern psychology—behaviorism, psychoanalysis, and transpersonal psychologies—with the view of the self that is expressed in the great religious traditions of the world.

Chapter 10

EXPLORING THE MIND-BODY PROBLEM

BEFORE WE INVESTIGATE how the idealist philosophy and quantum theory can be applied to the mind-body problem, let us review the prevailing contemporary philosophy. We all share an overwhelming intuition that our mind is separate from our body. There is also the conflicting intuition that mind and body are the same—as when we are in bodily pain. Additionally, we intuit that we have a self separate from the world, an individual self that is conscious of what is going on in our minds and bodies, a self that wills (freely?) some of the actions of the body. The philosophers of the mind-body problem examine these intuitions.

First, there are philosophers who posit that our intuition of a mind (and consciousness) separate from the body is right. These are the dualists. Others deny dualism; they are the monists. There are two schools of monists. One school, the material monists, feels that body is primary and that mind and consciousness are but epiphenomena of the body. The second school, the monistic idealists, posits the primacy of consciousness, with mind and body being epiphenomena of consciousness. In Western culture, particularly in recent times, the material monists have dominated the monist school. In the East, on the other hand, monistic idealism has remained a force.

There are many ways of thinking about the mind-body problem, many ways to reach conclusions, and many subtleties to be accounted for. As we embark on a tour of what I will call the

University of Mind-Body Studies, I would like you to bear these subtleties in mind. Imagine that all great mind-body thinkers are here, now, at the University of Mind-Body Studies, where the traditional faculty from throughout history teaches the solutions—old and new, dualist and monist—to the mind-body problem. Before you enter the university, a word of caution: Retain your skepticism and always refer any philosophy to your own experience before you invest your allegiance.

You find the university with little difficulty—there is an enticing aroma about it. Upon closer approach, you recognize the source of the aroma in a fountain named Meaning at the entrance. The elixir flowing from the fountain is ever changing, but its aroma is always alluring.

You pass through the gate and look around. The buildings are of two distinct styles. On one side of the street there is an old, very elegant structure. You have a weakness for classical architecture, so you turn that way. The modern highrise on the other side can wait.

As you approach the building, however, a picketer stops you and hands you a pamphlet that reads

BEWARE OF DUALISM

The dualists are taking advantage of your naiveté to teach outdated ideas. Consider this: Suppose one of the robots in a Japanese automobile factory were conscious and you asked for its opinion on the mind-body problem. According to our leader, Marvin Minsky, "When we ask such a creature what sort of being it is, it simply cannot answer directly; it must inspect its models. And it must answer by saying that it seems to be a dual thing—which appears to have two parts—a 'mind' and a 'body'."[1] Robot thinking is primitive thinking. Don't succumb to it. Insist on monism for solutions that are modern, scientific, and sophisticated.

"But," you object to the picketer, "I sometimes feel myself that way, as mind and body, separate. You are not saying . . . But who asked you, anyway! And for your information, I like old wisdom. I want to check it out myself, if you will move out of the way, please."

The picketer makes way for you with a shrug. In front of the building there is a signpost saying, Hall of Dualism, René Descartes, Dean. The very first office you enter engulfs you in nostalgia. A middle-aged man, a professor you assume, silently gazes at the ceiling. Somehow the familiarity of his face makes you feel that you should recognize him. Suddenly, you notice the insignia on his desk: *Cogito, ergo sum.* Of course! This must be René Descartes.

Descartes returns your greeting with a kind smile. His eyes shine as he responds in a dignified voice to your request for an explanation of the mind-body relation. His explanation of *cogito, ergo sum* is elegant: "I can doubt everything, even my body, but I cannot doubt that I think. I cannot doubt the existence of my thinking mind, but I can doubt the body. Obviously, mind and body must be different things." He says that there are two independent substances, soul substance and physical substance. Soul substance is indivisible. Mind and soul—the indivisible, irreducible part of reality that is responsible for our free will—are made of this soul substance. Physical substance, on the other hand, is infinitely divisible, reducible, and governed by scientific laws. But only faith governs the soul substance.

"Freedom of the will is self-evident," he says in answer to a question, "and only our mind can know that."

"Because our mind is independent of the body?" you ask.

"Yes."

But you are not satisfied. You remember that Cartesian dualism of mind and body violates the laws of conservation of energy and momentum that physics has established beyond doubt. How could mind possibly interact with the world without occasionally exchanging energy and momentum? But we always find the energy and momentum of objects in the physical world to be conserved, to remain exactly the same. As soon as you see an opportunity, you murmur your apology and get out of Descartes' office.

The next office has the name Gottfried Leibniz inscribed on it. As you enter, professor Leibniz affably inquires: "What were you doing in there with old Descartes? Everybody knows that the good Descartes' interactionism doesn't hold water. How can an immaterial soul be so materially localized in the pineal gland?"

"Do you have a better explanation?"

"Of course. We call it psychophysical parallelism." He summarizes: "Mental events run independent of but parallel to physiological events within the brain. No interaction, no embarrassing questions." He smiles complacently.

But you are disappointed. The philosophy does not explain your intuition that you have free will, that your self has causal power over the body. It sounds suspiciously like sweeping the dirt under the rug—out of sight, out of mind. As you are grinning to yourself at the private pun, you notice somebody beckoning you.

"I am Professor John Q. Monist. Your head must be spinning from all that dualistic talk about the mind," he says.

You admit a growing mental fatigue, and the man responds, sounding a little sarcastic: "Mind is the ghost in the machine." In response to your obvious puzzlement, he continues: "A visitor came to Oxford and was shown all the colleges, the buildings, and so forth. Afterward, he wanted to know where the university was. He didn't realize that the colleges are the university. The university is a ghost."

"I think mind must be something more than a ghost. After all, I do have self-consciousness . . ."

The man interrupts you. "It's all mirage; the problem is one of using improper language," he says testily. "Go to the monists on the other side. They will tell you."

Perhaps the man is right; the monists may be the professors of truth, after all. Certainly, there are many more offices in the huge, sleek building on the other side.

But there is a picketer there too. "Before you go in there," the picketer pleads, "I just want you to be aware that they will try to bamboozle you with promissory materialism; they will insist that you ought to accept their claims because 'surely' the proof is forthcoming." You promise to be careful, and he moves aside. "I'll keep my fingers crossed," he says, crossing his fingers.

The lobby is clamorous, but most of the noise seems to come from a lecture room where a poster identifies the subject of the lecture as Radical Behaviorism. Inside the hall, a man is pacing and lecturing behind a lectern to a rather sparse audience. As you approach nearer, you realize that the speaker is talking about the work of the famous behaviorist B. F. Skinner. Of course! The sign in front of the school indicated that Skinner is the dean, naturally his work would find some prominence here.

"According to Skinner, the mentalist problem can be avoided by going directly to the prior physical causes while bypassing intermediate feelings or states of mind," the speaker is saying. "Consider only those facts that can be objectively observed in the behavior of one person in its relation to his prior environmental history."[2]

"Skinner wants to dispense with the mind—no mind, no mind-body problem—the same way the parallelists try to eliminate the problem of interaction. To me they both smack more of running away from a problem than of solving it," you tell the professor next door.

"True, radical behaviorism is too narrow in scope. We should study the mind, but only as an epiphenomenon of the body. Epiphenomenalism," the professor explains, "is the idea—the only idea, by the way, that makes sense in the mind-body problem—that mind and consciousness are epiphenomena of the body, secreted by the brain as the liver secretes bile. Tell me, what else can it be?"

"It's your job to tell me; you are the philosopher. Explain how the epiphenomenon of self-consciousness arises from the brain."

"I haven't found out yet. But surely we will. It's only a matter of time," he insists, waggling his index finger.

"Promissory materialism, just as the picketer warned!" you mutter and leave.

In the office across the hall, Professor Identity is insistent.[3] He does not want you to leave his department without a whiff of the truth. To him, identity is truth—mind and brain are identical. They are two aspects of the same thing.

"But that doesn't explain my experiences of the mind; if that's all you have to say, I am not interested," you declare, edging toward the door.

But Professor Identity is insistent that you understand his position. He says that you must learn to replace mental terms in your language with neurophysiological terms because corresponding to every mental state there is ultimately a physiological state that is the real McCoy.

"Somebody else is preaching something like that—parallelism, he calls it." You feel really pleased that you are able now to toss off philosophical terms without faltering.

With practiced smoothness, Professor Identity gives another interpretation of identity theory: "Even though the mental and the physical are one, we distinguish between them because they represent different ways of knowing things. You have to learn the logic of categories before you fully understand this, but . . ."

That last pontification finally blows your top, and you sock it to him. "Look, I've been roaming from one office to another for hours now with a simple inquiry: What is the nature of our mind that gives it free will and consciousness? And all I hear is that I cannot have such a mind."

Identity is undaunted. He mutters something to the effect that consciousness is a woolly concept.

"Consciousness is woolly, huh?" You are still angry. "Are you and I woolly? Then why do you take yourself so seriously?"

You make your escape in a hurry, before the bewildered Identity has a chance to answer you. It is possible, you muse to yourself on your way out, that your action was a conditioned response initiated in your brain and simultaneously arising in your mind as what seemed like free will. Can one really know if one has free will by any philosophical trick, or is philosophy hopeless? But philosophy can wait, all you care about right now is some pizza and a tall glass of beer.

A dimly lit part of the building diverts your attention. On closer examination, you discover that this building has older architecture. The new building has been built on parts of it. There is a sign: "Idealism. Enter at your own risk. You may never again be a proper philosopher of the mind-body." But the warning only increases your curiosity.

The first office belongs to Professor George Berkeley. Interesting man, this Berkeley. "Look, any statements you make about physical things are ultimately about mental phenomena, perceptions, or sensations, aren't they?" says he.

"That's true," you answer, impressed.

"Suppose you wake up all of a sudden and find that you've been dreaming. How can you distinguish material stuff from dream stuff?"

"I probably can't," you admit. "There is, however, the continuity of experience."

"Continuity be darned. Ultimately, all you can trust, all you can be sure of, is mind stuff—thoughts, feelings, memories, and all that. So they must be the real."[4]

You like Berkeley's philosophy; it makes your free will real. Yet you are hesitant to call the physical world a dream. Besides, something else is bothering you.

"There doesn't seem to be any place in your philosophy for those objects that are not in anybody's mind," you complain.

But Berkeley is complacent: "Well, they are in God's mind."

And that sounds like dualism to you.

A semi-dark room catches your interest and you take a peek. Lo!

What's this? There is a shadow show on the wall projected by light from the back, but people watching the show are so strapped in their seats that they cannot turn. "What's going on?" you whisper to the woman with the light.

"Oh, this is Professor Plato's demonstration of monistic idealism. People see only the shadow show of matter and are beguiled by it. If only they knew that the shadows are cast by the 'realer' archetypal objects behind them, the ideas of consciousness! If only they had the fortitude to investigate the light of consciousness, which is the only reality," she laments.

"But what straps people to their seats, I mean in real life?" you want to know.

"Why do people like illusion better than reality? I don't know how to answer that. I know there are those in our faculty—Eastern mystics I think they are called—who say it's due to *maya*, which means illusion. But I don't know how *maya* works. Perhaps if you wait for the professor . . ."

But you don't wait. Outside, the hallway gets even dimmer, and an arrow is marked "To Eastern mysticism." You are curious, but you are also tired; you want beer and pizza. Maybe later. Surely, Eastern mystics will not mind waiting. Easterners are known for their patience.

But it's beer and pizza that have to wait. As you get outside the building, you get caught up in a big debate. A sign on one side says Mentalism, and you can't resist hearing these mentalists out. Who are the opponents? you wonder. There! A sign says Physicalism.

Presently, it is the physicalists who have the floor. The speaker seems rather sure of herself: "In the reductionistic view, mind is the higher level of a hierarchy of levels, and the brain, the neuronal substratum, is the lower level. The lower level is the causal determinant of the higher; it cannot be the other way around. As Jonathan Swift explained:

So, naturalists observe, a flea
Hath smaller fleas that on him prey;
And these have smaller still to bite 'em;
And so proceed ad infinitum.

The smaller fleas bite the bigger, but the bigger fleas never affect the behavior of the smaller fleas."

"Not so fast," cautions a mentalist, as it is now their turn. "According to Roger Sperry, our guru, mental forces do not violate, disturb, or intervene in neuronal activities, but they do supervene; mental actions with their own causal logic take place as something additional to lower-level brain actions. The causally potent reality of the conscious mind is a new emergent order that arises from the organizational interaction of the neuronal substrata, but is not reducible to it."

The speaker pauses momentarily; a physicalist from the other side tries to butt in but to no avail: "Sperry holds subjective mental phenomena to be primary, causally potent realities as they are experienced subjectively, different from, more than, and not reducible to their physicochemical elements. The mental entities transcend the physiological just as the physiological transcend the molecular; the molecular, the atomic and subatomic; and so forth."[5]

The physicalist debater replies that such reasoning as Sperry's is all hocus-pocus, that what any conglomerate or configuration of neurons does is inevitably reducible to what the component neurons do. Every so-called causal action of the mind ultimately must be traceable to some underlying neuronal components of the brain. Mind initiating changes in the lower level of the brain is tantamount to having brain substratum acting on brain substratum without a cause. And where does the causal potency of the mind, free choice, come from? "Dr. Sperry's whole thesis is built on the unprovable theorem of holism—that the whole is greater than its parts. I rest my case." The speaker sits down complacently.

But the mentalists are ready with their rebuttal. "Sperry says that free will is that aspect of mental phenomena that is more than their physicochemical elements. Somehow this causally potent mind emerges from the interaction of its parts, of the myriad neurons. Clearly, the whole is greater than the parts. We just have to discover how."

The opposition is not ready to yield. Somebody with a big button that says *Think Functionalism* takes the podium. "We functionalists look at the brain-mind as a biocomputer, at brain as structure or hardware, and at mind as function or software. As you mentalists surely will agree, O ye misguided supporters of mentalism, the computer is the most versatile metaphor ever invented to describe the brain-mind. And as you know, we don't completely accept the reductionist view. Mental states and processes are functional entities implementable in different types of structure, be it the brain or the silicon computer. We can prove our point by building an artificial intelligence machine with mind—the Turing machine. But here again, although we use software language to describe mental processes as programs acting on programs, ultimately we know that all is the play of *some* hardware."[6]

"But there must be high-level programs of the mind that can initiate actions at the hardware level . . ." a mentalist tries to interrupt, but *Think Functionalism* does not yield.

"Your so-called high-level program, any program, is always implemented as hardware! So you have a causal circle, hardware acting on hardware without a cause. That is impossible. Your holism is nothing but dualistic thinking in disguise."

You see the mentalist getting agitated. It must be the ultimate insult for a mentalist to be called a dualist. But somebody is trying to divert your attention. "You are wasting your time. The physicalists are right. Mentalist thinking is pseudomonism; indeed, it does smack of dualism, but Sperry is also right. Mind does have supervention powers. The solution is a modern form of dualism. It's brand new. Here's the philosopher Sir John Dual. He will explain it to you."

As Dual begins to speak, you have to admit that he has charisma. "According to the model that Sir John Eccles and Sir Karl Popper have developed, mental properties belong to a separate world, world 2, and meaning comes from a still higher world, world 3.[7] Eccles says that a liaison brain located in the dominant cerebral hemisphere mediates between the brain states of world 1 and the mental states of world 2. Look, how can you deny that the capacity for creative freedom requires a jump out of the system. If you are all the system there is, your behavior is bound to be determined because any proposal of action-initiating mind is bound to end up in the paradoxical causal loop, brain—mind—brain, that snared Sperry."

You are quite dazzled by Dual's charisma, or is it simply the accent? But what about the conservation laws? And doesn't Eccles's liaison brain sound like another form of the pineal gland? It does to you. But lo! before you ask these questions something else attracts your attention—a sign, The Chinese Room, adjacent to a closed box with a couple of openings.

"This is a debunking device, built by professor John Searle of U.C. Berkeley, that shows the inadequacy of the functionalist,

Turing machine view of the mind.[8] I'll explain how it works in a minute," says an amicable guy. "But suppose you get into the box first."

You are a little surprised, but you comply. You are not passing up a chance to experience the debunking of the Turing machine. Soon a flash card comes at you through a slot. On the card are scribbled some characters that you suspect are Chinese, but not knowing Chinese, you don't recognize their meaning. There is a sign in English telling you to consult a dictionary, also in English, where an instruction is given for the response card that you have to find from a pile of cards. After some effort, you find the response card and present it to the outgoing slot as instructed.

When you come out, you are greeted with smiles. "Did you understand the semantic situation at all? Do you have any idea what meaning was conveyed by the cards?"

"Of course not," you say, a little impatiently. "I don't know Chinese if that's what it was, and I am not clairvoyant."

"Yet you were able to process the symbols just as a Turing machine does!"

You catch on. "So, like me, the Turing machine need not have any understanding of what communication goes on when it processes symbols. Just because it manipulates symbols, we cannot be sure that it also understands."

"And if the machine cannot understand when it processes symbols, how can we say that it thinks?" says the man who speaks for John Searle.

You have to admire Searle's ingenuity. But then if the functionalists' claim is wrong, their picture of the mind-body relation must be wrong, too. Sperry's idea of emergence is akin to dualism. And dualism is dubious even when sold in the new Popper bottle. Is there any way to understand consciousness and free will? you wonder. Maybe old Skinner is right—we should just analyze behavior and be done with it.

What is all that commotion near the fountain yonder? You do not expect an East Indian Buddhist monk on a chariot to argue with somebody who could only be a king—throne, crown, and all. To your amazement, the monk begins to strip his chariot. First he removes the horses from the chariot and asks, "Are these horses the chariot, O noble king?"

The king replies, "Of course not."

The monk then removes the wheels and asks, "Are the wheels the chariot, O noble king?"

Receiving the same reply, the monk continues the process until all the detachable parts of the chariot have been removed. He then points to the chassis of the chariot, asking for the last time, "Is this the chariot, O noble king?"

Once again the king replies, "Of course not."

You can see the irritation in the king's face. But of course, to you, the monk has made his point. Where is the chariot?

You should have had lunch because you are feeling lightheaded as the exotic images flash before you. Then, like magic, Professor John Q. Monist appears before you again and says scornfully: "See, I told you. There is no chariot without the reductive parts. The parts are the whole. Any concept of chariot apart from the parts is a ghost in the machine."

And now you become really puzzled, beer and pizza forgotten completely. How can a Buddhist monk—a bona fide Eastern mystic who is supposed to belong to the idealist camp—make arguments that give ammunition to somebody as cynical as Professor Monist?

There is no puzzle here if you are familiar with Buddhism. The Buddhist monk (his name was Nagasena, and the king is King Millinda) may sound the same as Professor Monist, since they both deny self-nature to objects. However, according to material monists, there is no self-nature in objects apart from the ultimate reductive components, the elementary particles that make them up. Nagasena's position—monistic idealism—is radically different. There is no self-nature in objects apart from consciousness.

Take special note that there is no need to ascribe self-nature to subjects either. (This is where Berkeley's kind of idealism finds criticism.) In vintage monistic idealism, only transcendent and unitive consciousness is real. The rest, including the subject-object division of the world, is epiphenomenon, maya, illusion. This is philosophically astute but not completely satisfactory. The doctrine of no-self (or the illusory nature of the self) does not explain how the individual self-experience arises. It does not explain our very personal "I." Thus one of our most compelling experiences is left out.

So that is our brief review of philosophy. Dualism has difficulty in explaining mind-body interaction. Material monists negate free will and hold consciousness to be an epiphenomenon, merely the software clamoring of our biocomputer hardware. Even monistic idealists fall short because they too undermine the experience of the personal self, being too enamored by the whole. Can quantum mechanics break the deadlock on some of these tough questions?

Chapter 11

IN SEARCH OF THE QUANTUM MIND

IN THE LAST CHAPTER we saw that none of the philosophical answers to the mind-body problem is completely satisfactory. The most satisfactory philosophy seems to be monistic idealism because it is based on consciousness being the primary reality, but even monistic idealism leaves unanswered the question of how our individual, personal "I" experience emerges.

Why is personal selfhood a difficult problem for idealism? Because in idealism consciousness is transcendent and unitive. One might well ask why, then, and how the sense of separateness arises? A traditional answer given by such idealists as Shankara is that the individual self is, like the rest of the immanent world, illusory. It is part of what is called in Sanskrit *maya*, the world illusion. In a similar vein, Plato called the world a shadow show. But no idealist philosopher ever explains why such an illusion exists. Some flatly deny that an explanation can ever be found: "The doctrine of *maya* recognizes the reality of multiplicity from the relative standpoint (of the subject-object world)—and simply states that the relationship of this relative reality with the Absolute (undifferentiated, unmanifest consciousness) cannot be described or known."[1] This is an unsatisfactory response. We want to know whether the individual "I" experience really is an illusion, an epiphenomenon. If it is, we want to know what creates the illusion.

If you saw an optical illusion, you would immediately seek an explanation, wouldn't you? This individual "I" experience is the most persistent experience of our lives. Should we not seek an explanation of why it arises? Maybe if we find out how the individual "I" arises, we will be able to understand ourselves better. Can we explain *maya* with our model? In this chapter I shall present a view of the mind and brain (a system we can call the brain-mind) that accounts, within the framework of monistic idealism, for our individual, separate self-experience.

IDEALISM AND THE QUANTUM BRAIN-MIND

In the past few years it has become increasingly clear to me that the only view of the brain-mind that is complete and consistent in its explanatory power is this: The brain-mind is an interactive system with both classical and quantum components. These components interact within a basic idealist framework in which consciousness is primary. In this and the next couple of chapters, I shall examine the solution of the mind-body problem offered by such a view. I shall show that this view, unlike other solutions to the mind-body problem, accounts for consciousness, cause-effect relations in matters of the brain-mind (that is, the nature of free will), and the experience of personal self-identity. Additionally, we shall find that this solution reveals creativity to be a fundamental ingredient of human experience.

The distinction of quantum and classical machinery in the present answer is, of course, purely functional (in the sense described in chapter 9). *The quantum component of the brain-mind is regenerative and its states are multifaceted. It is the vehicle for conscious choice and for creativity*. In contrast, because it has a long regeneration time, *the classical component of the brain-mind can form memory and thus can act as a reference point for experience*.

You may ask, Is there any evidence at all that the ideas of quantum mechanics apply to the brain-mind? There seems to be at least circumstantial evidence.

David Bohm and before him August Comte noted that there seems to be an uncertainty principle operating for thought.[2] If we concentrate on the content of thought, we lose sight of the direction in which the thought is heading. If we concentrate on the direction of a thought, we lose sharpness in its content. Observe your thoughts and see for yourself.

We can generalize Bohm's observation and posit that thought has an archetypal component. Its appearance in the field of awareness is associated with two conjugate variables: *feature* (instantaneous content, akin to the position of physical objects) and *association* (the movement of thought in awareness, akin to the momentum of physical objects). Note that awareness itself is akin to the space in which thought objects appear.

So, mental phenomena such as thought seem to exhibit complementarity. We can posit that, although it is always manifested in form (described by attributes such as feature and association), between manifestations thought exists as transcendent archetypes—as does the quantum object with its transcendent coherent superposition (wave) and manifest one-faceted (particle) aspects.

Additionally, there is plenty of evidence of discontinuity—quantum jumps—in mental phenomena, especially in the phenomenon of creativity.[3] Here is a compelling quote from my favorite composer, Tchaikowsky: "Generally speaking, the germ of a future composition comes suddenly and unexpectedly. . . . It takes root with extraordinary force and rapidity, shoots up through the earth, puts forth branches and leaves, and finally blossoms. I cannot define the creative process in any way but [by] this simile."[4]

This simile is exactly the kind that a quantum physicist might use to describe a quantum leap. I shall spare you further quotes, but great mathematicians, such as Jules-Henri Poincaré[5] and Carl Friedrich Gauss,[6] have spoken of their own creative experiences in similar terms, as being sudden and discontinuous like a quantum leap.

A Sidney Harris cartoon makes the same point quite well. Einstein, baggy pants and all, stands before a blackboard with chalk in hand, ready to discover a new law. On the board, the equation $E = ma^2$ is written and crossed out. Under this, $E = mb^2$ is also written and crossed out. The caption reads, "The Creative Moment." Is $E = mc^2$ going to burst forth? Not likely. The cartoon is a caricature of a creative moment precisely because we all intuitively recognize that the creative moment does not follow such continuous, reasoned steps. (For an excellent treatment of the so-called sloppiness and lack of rigor of the actual business of doing mathematics, see George Polya's delightful book *How to Solve It*.)

There is evidence of nonlocality in the mind's action as well, not only in the controversial distant-viewing data cited previously but also in recent brainwave coherence experiments that we will discuss later in this book.

Tony Marcel's research supports the idea of the quantum component of the brain-mind. These data are important enough to deserve special attention.

TONY MARCEL'S DATA REVISITED

For more than a decade, Tony Marcel's data have eluded a fully satisfactory explanation by existing cognitive models. These data involve measuring the recognition time for the last word in three-word strings, such as *tree-palm-wrist* and *hand-palm-wrist*, in which the middle ambiguous word is sometimes pattern-masked so that it can be perceived only unconsciously.[7] The effect of pattern masking seems to be to remove the congruent (as in the case of *hand*) or incongruent (as in the case of *tree*) effect of the first (priming) word on the recognition time.

The no-mask situation, in which the subjects are aware of the second word, supports what is called the selective theory of the effect of prior context in word recognition.[8] Word one affects the perceived meaning of the polysemous word, word two. Only the biased meaning of word two (biased by the effect of word one) is passed on. If this meaning is congruent (incongruent) with the target word, we get facilitation (inhibition) of recognition—short (long) recognition time. If the brain-mind is looked upon as a classical computer, as in functionalism, then the computer seems to operate in a serial, top-down, linear, and unidirectional fashion in this kind of situation.

When the polysemous word is pattern-masked, both of its meanings seem to be available in the subsequent processing of information—regardless of the presence of biasing context—since the congruent and incongruent conditions take similar recognition times. Marcel himself cited the importance of distinguishing between conscious and unconscious perceptions and noted that a nonselective theory must apply to the unconscious identification. (The selective theory applies only to conscious perception.) In addition, it appears that such a nonselective theory must be based on parallel processing, in which multiple units of information are simultaneously processed with feedback included.[9] Such parallel-distributed processing models are examples of the bottom-up, connectionist approach to artificial intelligence machines, in which the connections among the various components play a dominant role.

Without going into too many technical details, classical functionalist models that are linear and selective have no difficulty in explaining the effect of biasing the context in cases where no masks are used, but these models cannot explain the significant change that occurs in the unconscious perception experiment with pattern masking. The same is true for theories of nonselective parallel processing. They can be adjusted to fit either piece of the data—the case of conscious perception or the case of unconscious perception—but both sets cannot be explained in a coherent fashion. Hence, concludes Marcel in the work previously cited, "these [masking] data are inconsistent with and qualitatively different from those in the no-masking condition." Thus the distinction of conscious and unconscious perception in the Marcel data has been a problem for supporters of cognitive models.

The psychologist Michael Posner has a cognitive solution that invokes attention as the crucial ingredient for distinction between conscious and unconscious perception.[10] Attention comes with selectivity. Thus, according to Posner, we select one of two meanings when we are attentive, as in conscious perception of the ambiguous word in the Marcel experiment. When we are not attentive, there is no selection. Thus both meanings of an ambiguous word are perceived, as in the unconscious perception of the pattern-masked word in Marcel's experiment.

So who turns attention on or off? According to Posner, a central processing unit switches attention on or off. Nobody, however, has ever found a central processing unit in the brain-mind, and the concept raises the specter of the so-called little human, or homunculus, inside the brain. The Nobel laureate biologist Francis Crick alludes to the problem in the following anecdote: "Recently I was trying to explain to an intelligent woman the problem of understanding how it is we perceive anything at all, and I was not having any success. She could not see why there was a problem. Finally in despair I asked her how she herself thought she saw the world. She replied that she probably had somewhere in her head something like a television set. 'So who,' I asked, 'is looking at it?' She now saw this problem immediately."[11]

We may as well face it: There is no local homunculus, or central processing unit, sitting in the brain that switches attention, that interprets and ascribes meaning to all the actions of the mental conglomerates, tuning the channels from a control room. Thus, self-reference—our ability to refer to our "I" as the subject of our experiences—is a very difficult problem for the classical functionalist models, top-down or bottom-up. What we are looking for is what is looking—this essential reflexivity is as difficult to explain in materialist models of the brain-mind as the von Neumann chain is in quantum measurement.

Suppose, however, that when somebody sees a pattern-masked word having two possible meanings, the brain-mind becomes a quantum coherent superposition of states—each carrying one of the two meanings of the word. This assumption can explain both sets of Marcel data—conscious and unconscious perception—without invoking a central processing unit.

The quantum mechanical interpretation of the conscious perception data is that the contextual word *hand* projects out of the dichotomous word *palm* (a coherent superposition) the state with the meaning of hand (that is, the wave function collapses with the choice of hand meaning only). This state has a large overlap (positive associations are expressed in quantum mechanics as large overlaps of meaning between two states) with the state corresponding to the final word *wrist*, and thus the recognition of wrist is facilitated.

Similarly, in the quantum model description of the unmasked incongruent case, the word context of *tree* projects out the state with the meaning of tree from the coherent superposition state *palm*; the overlap of meaning between the states corresponding to *tree* and *wrist* is small, hence the inhibition. In the pattern-masked case, both congruent and incongruent, *palm* is perceived unconsciously and therefore there is no projection of any particular meaning—no collapse of the coherent superposition. Direct evidence of pattern-masked *palm* leading to a coherent superposition state containing both the tree and hand meanings of *palm* can thus be seen. How else would the effect of the biasing word, as in the strings *tree-palm-wrist*/*hand-palm-wrist*, get so nearly wiped out when *palm* is pattern-masked?

The phenomenon of simultaneously accessing *palm* as both a tree and a part of the hand is difficult to account for accurately in a classical linear description of the brain-mind because such a description is either/or. The advantage of the "both-and" quantum description is obvious.[12]

I realize that the data suggesting the parallels between the mind and the quantum—uncertainty, complementarity, quantum jumps, nonlocality, and finally, coherent superposition—may not be considered conclusive. They could well be indicative, however, of something radical: *What we call the mind consists of objects that are akin to the objects of submicroscopic matter and that obey rules similar to those of quantum mechanics.*

Let me put this revolutionary idea differently. Just as ordinary matter consists ultimately of submicroscopic quantum objects that can be called the archetypes of matter, let us assume that the mind consists ultimately of the archetypes of mental objects (very much like what Plato called ideas). I further suggest that they are made of the same basic substance that material archetypes are made of and that they also obey quantum mechanics. Thus quantum-measurement considerations apply to them as well.

QUANTUM FUNCTIONALISM

I am not alone in this kind of speculation. Jung intuited decades ago that psyche and matter must ultimately be made of the same stuff. In recent years several scientists have seriously attempted to invoke a quantum mechanism in the macroscopic working of the brain-mind to explain brain data. What follows is a brief summary of their efforts.

How does an electrical impulse pass from one neuron to another across a synaptic cleft (the place where one neuron feeds into another)? Conventional theory says that the synaptic transmission must be due to a chemical change. The evidence for this is somewhat circumstantial, however, and E. Harris Walker has challenged it in favor of a quantum-mechanical process.[13] Walker thinks that the synaptic cleft is so small that the quantum tunneling effect may play a crucial role in the transmission of nerve signals. Quantum tunneling is the ability that a quantum object has to pass through an otherwise impassable barrier, an ability arising from its wave nature. John Eccles has discussed a similar mechanism for invoking the quantum in the brain.[14]

The Australian physicist L. Bass and, more recently, the American Fred Alan Wolf have observed that for intelligence to operate, the firing of one neuron must be accompanied by the firing of many correlated neurons at macroscopic distances—as much as ten centimeters, which is the width of the cortical tissue. In order for this to happen, notes Wolf, we need nonlocal correlations (in the manner of Einstein, Podolsky and Rosen, of course) existing at the molecular level in our brain, at our synapses. Thus even our ordinary thinking depends on the nature of quantum events.[15]

The Princeton scientists Robert Jahn and Brenda Dunn have used quantum mechanics as a model for the paranormal abilities of the brain-mind, if only as a metaphor.[16]

Consider once again the model that the functionalists use—that of the classical computers. Richard Feynman once proved mathematically that a classical computer can never simulate nonlocality.[17] Thus the functionalists are forced to deny the validity of our nonlocal experiences, such as ESP and meaning, because their model of the brain-mind is based on the classical computer (which is incapable of instantiating or modeling nonlocal phenomena). What gigantic shortsightedness! To borrow Abraham Maslow's phrase once again: If you have only a hammer, you treat everything as if it were a nail.

Can one, however, simulate consciousness without nonlocality? I am speaking about consciousness as we humans experience it—a consciousness that is capable of creativity, of love, of freedom of choice, of ESP, of the mystical experience—a consciousness that dares to form a meaningful and evolving worldview in order to understand its place in the universe.

Perhaps the brain accommodates consciousness because it has a quantum system sharing the job with its classical one, say the University of Alberta biologist C. I. J. M. Stuart and his collaborators, physicists M. Umezawa and Y. Takahashy,[18] and the Berkeley physicist Henry Stapp.[19] In this model, which I have adapted (see the next section), the brain-mind is looked upon as two interacting classical and quantum systems.[20] The classical system is a computer that runs on programs that for all practical purposes follow the deterministic laws of classical physics and, therefore, can be simulated in algorithmic form. However, the quantum system runs on programs that are only partly algorithmic. The wave function evolves according to the probabilistic laws of the new physics—this part is algorithmic, continuous. There is also the discontinuity of the collapse of the wave function, which is fundamentally nonalgorithmic. Only the quantum system displays quantum coherence, a nonlocal correlation among its components. Also, the quantum system is regenerative and thus can handle the new (because quantum objects remain forever new). The classical system is necessary to form memories, to make records of collapsed events, and to create a sense of continuity.

The marshaling of suggestive ideas and data can go on, but the point is simple: The conviction has been growing among many physicists that the brain is an interactive system with a quantum mechanical macrostructure as an important complement to the classical neuronal assembly. Such an idea is not yet a bandwagon by any means, but neither is it a lonely oxcart.

The Brain-Mind as Both Quantum System and Measuring Apparatus

We look at the brain-mind's quantum system technically as a macro quantum system consisting of many components that not only interact via local interactions but are also correlated in the manner of EPR. How does one represent the states of such a system?

Imagine two pendulums hung from a taut string. Better yet, imagine you and your loved one swinging as the pendulums. The two of you now make a system of coupled pendulums. If you set yourself in motion, but your loved one is still, very soon your loved one will start swinging as well—so much so that in no time he or she will take up all the energy and you will come to rest. Then the cycle will repeat. Something is lacking, though. There is not much togetherness. To correct the problem, you can both start swinging simultaneously in the same phase. Starting in this way, you will move together in a motion that would go on forever if there were no friction. The same would be true if you were to begin swinging together in opposite phase. These two ways of swinging are called the normal modes of the double pendulum. (The correlation between the two of you, however, is quite local, made possible by the taut string that supports your pendulums.)

We can similarly represent the states of a complex system, albeit quantum, by its so-called normal modes of excitation, its quanta, or more generally, by conglomerates of the normal modes. (It is too early to name these mental quanta, but at a recent conference on consciousness that I attended, we had fun playing around with such names as psychons, mentons, and so forth.)

Suppose these normal modes constitute the mental archetypes that I mentioned earlier? Jung found that mental archetypes have a universal character; they are independent of race, history, culture, and geographical origin.[21] This fits in rather nicely with the idea that Jungian archetypes are conglomerates of universal quanta—

the so-called normal modes. I will call the states of the brain's quantum system that are made up of these quanta *pure mental states.* This formal nomenclature will be useful later in our discussion.

Suppose also that the bulk of the brain is the classical analog of the measurement apparatus that we use to amplify submicroscopic material objects in order to see them. Suppose that the classical apparatus of the brain amplifies and records the quantum mind objects.

This solves one of the most persistent riddles of the brain-mind problem—the problem of brain-mind identity. Currently, philosophers either postulate brain-mind identity without clarifying what is identical with what or try to define some kind of psychophysical parallelism. For example, in classical functionalism, we can never truly establish the relation of mental states and the states of the computer.

In the quantum model, the mental states are states of the quantum system and, with a measurement, these states of the quantum brain become correlated with the states of the measuring apparatus (just as the state of the cat becomes correlated with the state of the radioactive atom in the Schrödinger's cat paradox). Thus in every quantum event the brain-mind state that is collapsed and experienced represents a pure mental state that the classical brain measures (amplifies and records), and there is a clear definition of and justification for the identity.

The recognition that most of the brain is a measuring apparatus leads to a new and useful way to think about the brain and conscious events. Biologists often argue that consciousness must be an epiphenomenon of the brain because changing the brain by natural damage or drugs changes conscious events. Yes! says the quantum theorist, because changing the measurement apparatus does certainly change what can be measured, and therefore, change the event.

The idea that the formal structure of quantum mechanics should apply to the brain-mind is not new at all but has been evolving gradually. However, the idea of looking at the brain-mind as a quantum system/measuring apparatus is new, and it is the consequences of this hypothesis that I want to explore here.

The brain scientists who have a materialist ax to grind will object. Macroscopic objects, objects in bulk, obey classical laws, albeit approximately. How could a quantum mechanism apply to the brain's macrostructure enough to make a difference?

Those of us who want to explore consciousness will overrule the objection. There are some exceptions to the general rule that objects in the macrocosm obey classical physics, even approximately. A few systems exist that cannot be explained by classical physics even at the macro level. One such system that we have already discussed is the superconductor. Another familiar case of a quantum phenomenon at the macro level is the laser.

A laser beam travels to the moon and back while maintaining its form as a narrow pencil beam because the photons of its beam exist in coherent synchrony. Have you ever watched people dance without music? They will be quite out of beat with each other, right? But tap out a rhythm, and they will be able to dance in perfect consort. The coherence of the photons of the laser beam arises from the beat of their quantum-mechanical interactions operating even at the macro level.

Could it be that a quantum mechanism in our brain, operating in ways similar to the laser,[22] opens itself to the supervention of nonlocal consciousness, with the classical parts of the brain performing the role of measuring apparatus for amplifying and making (if only temporary) records? I am convinced that the answer is yes.

Does the kind of coherence that the laser exhibits exist between different brain areas in certain mental actions? Some direct evidence of such coherence has indeed been found. Researchers into meditation have studied brain waves from different parts of the brain, front and back or left and right, to see if they exhibit any similarity in phase.[23] Using sophisticated techniques, these researchers have shown coherence in brain waves from different parts of the scalps of subjects in meditative states. The initial reports of spatial coherence of brain waves during meditation have since been confirmed by other researchers. Furthermore, the degree of coherence is found to be directly proportional to the degree of pure awareness that the meditator reports.

Spatial coherence is one of the startling properties of quantum systems. Thus these experiments on coherence may be giving us direct evidence that the brain acts as a measuring apparatus for the normal modes of a quantum system, which we may call the *quantum mind.*

More recently, the electroencephalogram (EEG) coherence experiment with meditative subjects has been extended to measuring brain-wave coherence on two subjects at once—with a positive result.[24] This is new evidence of quantum nonlocality. Two people meditate together, or are correlated via distant viewing, and their brain waves show coherence. Shouldn't even skeptics be intrigued? What else but EPR correlation between minds can explain such data.

The most definitive experimental support so far for the quantum in the brain-mind has come from the direct observation of EPR correlation between two brains by Jacobo Grinberg-Zylberbaum and his collaborators (chapter 8). In this experiment, two subjects interact for a period until they feel that a direct (nonlocal) connection has been established. The subjects then maintain their direct contact from within individual Faraday cages at a distance. When the brain of one of the subjects responds to an external stimulus with an evoked potential, the other subject's brain shows a transfer potential similar in form and strength to the evoked potential. This can be interpreted only as an example of quantum nonlocality due to the quantum nonlocal correlation between the two brain-minds established through their nonlocal consciousness.[25]

If the quantum computer sounds like Eccles's liaison brain and thus dualistic, do not worry. The quantum computer consists of quantum cooperation among some as yet unknown brain substrata. It is not a localized part of the brain, as the liaison brain is supposed to be, nor is its consciousness connection one that violates conservation of energy. Before the supervention of consciousness, the brain-mind exists as formless potentia (like any other object) in the transcendent domain of consciousness. When nonlocal consciousness collapses the brain-mind's wave function, it does so by choice and recognition, not by any energetic process.

What about the concern that the quantum brain is promissory, not an observed fact? It is true that the quantum brain-mind is a hypothesis. However, the hypothesis is based on solid philosophical and theoretical ground and is supported by plenty of suggestive experimental evidence. (The theory of blood circulation was formulated before the final piece of the puzzle, the network of capillaries, was discovered. Similarly, for the manifestation and circulation of mental processes in the brain, we need an EPR-correlated quantum network. It has to be there.) Moreover, the hypothesis is concrete enough to allow further theoretical predictions that can be subjected to experimental verification.[26] Additionally, because this theory recovers the classical (behavioral) limit as a new correspondence principle (which is explored in chapter 13), it is consistent with all the data that the old theory explains.

All new scientific paradigms start with hypotheses and theorizing. It is when the philosophy does not help formulate new theories and experimental tests or when it avoids confronting old, unexplained experimental data that it becomes promissory (as is material realism with regard to the problem of consciousness).

Bohr pointed out a complementarity principle between life and nonlife—the impossibility of studying life separately from the living organism—that may apply here.[27] The dual quantum system/classical measurement apparatus is a strongly interacting system, and it is this strong interaction, as we shall see, that is responsible for the appearance of individual and personal self-identity. It seems that there may be a complementarity here too. It may be impossible to study the brain's quantum system separately without destroying the conscious experience that is its trademark.

In summary, I have proposed a new way of looking at the brain-mind as both a measuring apparatus and a quantum system. Such a system involves consciousness as the collapser of the system's wave function, explains cause-effect relations as results of the free choices of consciousness, and suggests creativity as the new beginning that every collapse is. The groundwork follows for explaining how this theory accounts for the subject-object division of the world and eventually for the personal self.

Quantum Measurement in the Brain-Mind: A Partnership of the Classical and the Quantum

Classical functionalism assumes that the brain is hardware and the mind software. It would be just as unfounded to say that the brain is classical and the mind quantum. Instead, in the idealist model proposed here, the experienced mental states arise from the interaction of both classical and quantum systems.

Most important, the causal potency of the quantum system of the brain-mind arises from the nonlocal consciousness that collapses the mind's wave function and that experiences the outcome of this collapse. In idealism, the experiencer—the subject—is nonlocal and unitive; there is only one subject of experience. Objects appear from a transcendent possibility domain into the domain of manifestation when nonlocal, unitive consciousness collapses their wave functions, but we have argued that the collapse must occur in the presence of the awareness of a brain-mind in order for measurement to be completed. When we try to understand the manifestation of the brain-mind and awareness, however, we get into a causal circularity: There is no completion of measurement without awareness, but there is no awareness without the completion of measurement.

To see clearly both this causal circularity and the way out of it, we can apply quantum measurement theory to the brain-mind. According to von Neumann, the state of a quantum system undergoes change in two separate ways.[28] The first is a continuous change. The state spreads as a wave, becoming a coherent superposition of all the potential states allowed by the situation. Each potential state has a certain statistical weight given by its probability wave amplitude. A measurement introduces a second, discontinuous change in the state. All of a sudden the state of superposition, the multifaceted state that exists in potentia, is reduced to just one actualized facet. Think of the spreading out of the state of superposition as the development of a pool of possibilities, and think of the measurement process that manifests only one of the states of the pool (according to the probability rules) as a process of selection.

Many physicists view the selection process as random, an act of pure chance. This is the view that provoked Einstein's protest that God does not play dice. But if God does not play dice, who or what selects the result of a single quantum measurement? According to the idealist interpretation, it is consciousness that chooses—but a nonlocal unitive consciousness. The intervention of the nonlocal consciousness collapses the probability cloud of a quantum system. There is a complementarity here. In the manifest world, the selection process involved in the collapse appears to be random, while in the transcendent realm the selection process is seen as choice. As the anthropologist Gregory Bateson once remarked: The opposite of choice is random.

The brain-mind's quantum system must also develop in time, following the rules of measurement theory, and become a coherent superposition. The classical brain's functional machinery plays the role of the measuring apparatus and also becomes a superposition. Before the collapse, the state of the brain-mind thus exists as potentialities of myriad possible patterns that Heisenberg called tendencies. The collapse actualizes one of these tendencies, which leads to a conscious experience (with awareness) upon completion of the measurement. Importantly, the result of the measurement is a discontinuous event in space-time.

According to the idealist interpretation, consciousness chooses the outcome of the collapse of any and all quantum systems. This must include the quantum system that we postulate in the brain-mind. Thus we reach the consequence of talking about the interactive classical/quantum system of the brain-mind in the language of measurement theory as interpreted by monistic idealism: Our consciousness chooses the outcome of the collapse of the quantum state of our brain-mind. Since this outcome is a conscious experience, we choose our conscious experiences—yet remain unconscious of the underlying process. It is this unconsciousness that leads to the illusory separateness—the identity with the separate "I" of self-reference (rather than the "we" of unitive consciousness). The illusory separateness takes place in two stages, but the basic mechanism involved is called *tangled hierarchy*.[29] This mechanism is the subject of the next chapter.

Chapter 12

PARADOXES AND TANGLED HIERARCHIES

ONCE, when I was talking about tangled hierarchies, one of my listeners said that the phrase grabbed her interest even before she knew what it meant. She said that hierarchies reminded her of patriarchy and authority, but the term *tangled hierarchy* had a liberating tone to it. If your intuition is anything like hers, then you must be ready to explore the magical, perplexing world of language paradoxes and paradoxes of logic. Can logic be paradoxical? Is it not logic's forte to clear up paradoxes? The answers to these questions lead to tangled hierarchies.

As you approach the entrance to the catacomb of paradoxes, you encounter a creature of mythical proportions. You recognize her instantly as the Sphinx. Sphinx-like, she has a question for you, a question that you have to answer correctly to gain entrance: What creature is it that walks on four legs in the morning, two at midday, and three in the evening? Momentarily you are puzzled. What kind of question is this? Perhaps your trip will be nipped in the bud. You are only a beginner in this game of puzzles and paradoxes. Are you ready for what appears to be an advanced riddle?

To your great relief, along comes a Sherlock Holmes to aid your Mr. Watson. "I am Oedipus," he introduces himself. "The Sphinx's question is a riddle because it mixes up logical types, right?"

That's correct, you realize. It was useful to have learned about logical types before coming on this exploration. But now what? Fortunately, Oedipus continues. "Some of the words of the sentence have lexical meaning, but others have contextual meanings of higher logical type. It is the juxtaposition of the two types, typical of metaphors, that's causing your consternation." He gives you an encouraging smile.

Right, right. The words *morning*, *midday*, and *evening* must contextually refer to our lives—to our childhood, youth, and old age. Indeed, in our childhood we walk on four legs, but in our youth we walk on two, and three legs is a metaphor for two legs and a cane in our old age. It fits! You go over to the Sphinx and answer, "Man (or Woman)." The door opens.

As you walk through the door, something occurs to you. How did Oedipus, a mythical character from ancient Greece, know such modern terminology as *logical types*? But there isn't time to ponder: A new challenge demands your attention. A man pointing to another man at his side challenges you: "This man Epimenides is a Cretan who says, 'All Cretans are liars.'[1] Is he telling the truth or lying?" Well, let's see, you reason. If he is telling the truth, then all Cretans are liars, so he is lying—that's a contradiction. Okay, back to the beginning. If he is lying, then all Cretans are not liars and he may be telling the truth—that's also a contradiction. If you answer yes, the answer produces the reverberation of no, and if you answer no, then the reverberation comes back yes, ad infinitum. How can you solve such a puzzle?

"Well, if you can't solve the puzzle, at least you can learn to analyze it." As if by magic, another helper is at your side. "I am Gregory Bateson," he introduces himself. "What you are encountering is the famous liar's paradox: Epimenides is a Cretan who says, All Cretans are liars. The primary clause creates the context for the secondary clause. It classifies the latter. The secondary clause, if it were ordinary, would leave its primary clause alone, but no! This one reacts to reclassify the primary one, its own context."

"It's a mixture of logical types. I see it now," you brighten up.

"Yes, but it's not an ordinary mixture. Behold, the primary redefines the secondary. If yes, then no, then yes, then no. Forever it continues. Norbert Wiener used to say that if you fed this paradox to a computer, the computer would have had it. It would print out a series of Yes . . . No . . . Yes . . . No . . . Yes . . . until its ink ran out. It's a clever infinite loop that one cannot escape with logic."

"Isn't there any way of solving the paradox then?" You sound disappointed.

"Sure there is, because you are not a silicon computer," says Bateson. "I'll give you a hint. Suppose a salesman comes to your door with the following sales pitch: I have a beautiful hand fan for you for fifty bucks, which is a steal. Will it be cash or charge? What would you do?"

"I'd close the door in his face!" You know the answer to that one. (You remember the friend whose favorite game was a What-would-you-choose question—I chop off your hand, or I bite off your ear. That relationship ended in no time.)

"That's exactly right," Bateson smiles. "The way out of the infinite loop of a paradox is to slam the door, to jump out of the system. That gentleman over there has a good example." Bateson indicates a man who is seated at a table with a sign that reads "Only Two Can Play This Game."

The gentleman introduces himself as G. Spencer Brown. He claims that he does have a demonstration of how to get out of the game.[2] To get to it, however, you have to look at the liar's paradox in the form of a mathematical equation:

$$x = -1/x.$$

If you try the solution $+1$ on the right-hand side, the equation will give you back -1; try -1 and $+1$ comes back again. The solution oscillates between $+1$ and -1, just like the yes/no oscillation of the liar's paradox.

Yes, you can see that. "But what's the way out of this mad infinite oscillation?"

There is in mathematics a well-known solution of this problem, Brown tells you. Define the quantity called i as $\sqrt{-1}$. Notice that $i^2 = -1$. Dividing both sides of $i^2 = -1$ by i, you get

$$i = -1/i.$$

That is an alternative definition of i. Now try the solution $x = i$ on the left-hand side of the equation

$$x = -1/x.$$

The right-hand side now gives $-1/i$, which is equal to i by definition, no contradiction. Thus *i*, which is called an imaginary number, transcends the paradox.

"That's amazing." You are breathless. "You are a genius."

"It takes two to play the game," Brown winks.

Something in the distance attracts your attention: a tent with a big sign that reads "Gödel, Escher, Bach." As you approach the tent, a man with a boyish face manages to catch your eye and beckons brashly to you. "I am Dr. Geb," he says. "I spread the message of Douglas Hofstadter. I assume that you have read his book *Gödel, Escher, Bach*."[3]

"Yes," you mumble, somewhat taken aback, "but I didn't understand it completely."

"Look, it's really all very simple," says Hofstadter's messenger graciously. "All you need to understand is tangled hierarchies."

"Tangled what?"

"*Hierarchies*, not *what*, my friend. In a simple hierarchy the lower level feeds the upper level, and the upper level does not react back. In a simple feedback the upper level reacts back, but you still can tell what is what. With tangled hierarchies, the two levels are so thoroughly mixed that you cannot identify the different logical levels."

"But that's just a label," you shrug nonchalantly, still reluctant to accommodate Hofstadter's idea.

"You are not thinking. You have missed a very important aspect of tangled-hierarchical systems. I have been following your progress, you know."

"I assume that, in your wisdom, you will explain what I am missing," you say dryly.

"These systems—the liar's paradox is a prime example—are autonomous. They talk about themselves. Compare them with an ordinary sentence, such as Your face is red. An ordinary sentence refers to something outside itself. But the complex sentence of the liar's paradox refers to itself. That's how you get caught in its infinite delusion."

You hate to admit it, but this is a worthwhile insight.

"In other words," Hofstadter's messenger continues, "we are dealing with self-referential systems. The tangled hierarchy is a way of achieving self-reference."

"Dr. Geb, this is most interesting. I do have a certain interest in the matter of the self, so please tell me more," you capitulate. The man who spreads Hofstadter's message is not unwilling to oblige.

"The self arises because of a veil, a clear stonewalling against our attempt to see through the system logically. It is the discontinuity—in the liar's paradox, it is an infinite oscillation—that prevents us from seeing through the veil."

"I don't know if I am getting this."

Instead of explaining it once more, the Hofstadter enthusiast insists that you view a painting by the Dutch artist M. C. Escher. "In the Escher museum over there in that tent," he says, leading you toward it. "The name of the drawing is *Print Gallery.* It is most strange, but very much to the point of our discussion."

Inside the tent, you study the drawing (fig. 32). In the drawing, a

Figure 32. Escher's *Print Gallery,* a tangled hierarchy. The white spot in the middle indicates a discontinuity. © 1956 M. C. Escher/Cordon Art-Baarn-Holland. (Reprinted with permission from Escher Foundation.)

young man inside a gallery is looking at a picture of a ship that is anchored in the harbor of a town. But what's this? The town has a print gallery in which there is a young man who is looking at a ship that is anchored——

My God, it's a tangled hierarchy, you exclaim to yourself. After going through all these buildings of the town, the picture comes back to the original point where it starts, to begin its oscillation again and thus perpetuate the attention of the viewer to itself.

You turn to your guide with elation.

"You see the point." Your guide is all smiles.

"Yes, thank you."

"Did you notice the white spot in the middle of the drawing?" Dr. Geb suddenly asks. You did see it but did not pay much attention to it, you admit.

"The white spot, which has Escher's signature on it, shows how clear he was about tangled hierarchies. See, Escher could not have folded the picture back onto itself, so to speak, without violating the conventional rules of drawing, so there has to be a discontinuity. The white spot is the reminder to the observer of the discontinuity that is inherent in all tangled hierarchies."

"Out of the discontinuity comes the veil, and self-reference," you cry.

"Right." Dr. Geb is pleased. "But there is one more thing, one other aspect that you see best by considering the one-step self-referential sentence, I am a liar.[4] This sentence says that it is lying. This is the same system as the liar's paradox, which you encountered before—only the incidental clause-within-a-clause form has been eliminated. Do you see?"

"Yes."

"But in this form, something else begins to become clear. The self-reference of the sentence, the fact that the sentence is talking about itself, is not necessarily self-evident. For example, if you show the sentence to a child or a foreigner who is not very conversant with the English language, the response might be, Why are you a liar? He or she may not see at first that the sentence is referring to itself. Thus the self-reference of the sentence arises from our implicit, not explicit, knowledge of the English language. It is as if the sentence is the tip of an iceberg. There is a vast structure underneath that is invisible. We call it the inviolate level. It's inviolate from the system's point of view, of course. Take a look at another of Escher's drawings, this one named *Drawing Hands* (fig. 33)."

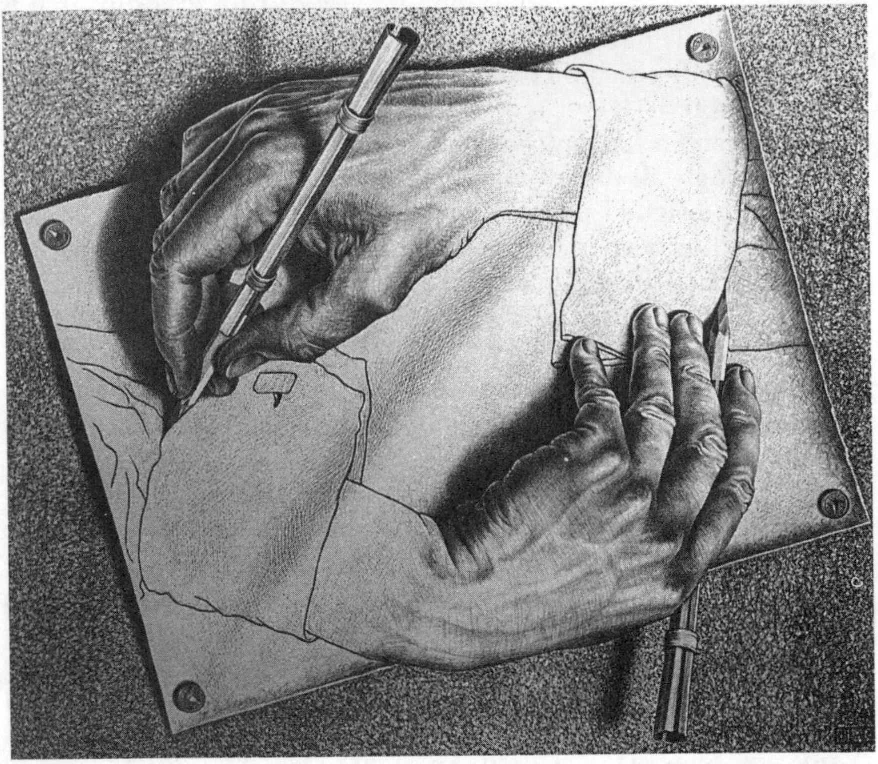

Figure 33. Drawing Hands, by M. C. Escher. © 1948 M. C. Escher/Cordon Art-Baarn-Holland. (Reprinted with permission.)

The left hand in this drawing is drawing the right hand, the right hand is also drawing the left; they are drawing each other. This is self-making, or autopoiesis. It is also a tangled hierarchy. And how is the system making itself? That particular illusion is created only if you stay within the system. From outside the system, from where you view it, you can see that the artist Escher has drawn both hands from the inviolate level.

Excitedly, you tell Dr. Geb what you see in Escher's picture. He nods in approval and says earnestly, "What interests Dr. Hofstadter about tangled hierarchies is this: He thinks that the programs of the brain's computer, the ones that we call the mind, form a tangled hierarchy, and out of this tangle comes our many-splendored self."

"Isn't that kind of a big jump?" You are suspicious of big jumps, always have been. One has to be cautious when wild-eyed scientists start making claims.

"Well, he has been thinking about the problem a lot, you know," Hofstadter's supporter says dreamily, "and I am sure that he will prove it someday by building a silicon computer with a conscious self."

You are impressed by Hofstadter's dream—our society needs people with dreams—but you feel a need to defend logic. "I must admit that I still am a little wary of tangled hierarchies," you say. "When I learned about logical types, I was told that they were invented to keep logic pure. But you, that is, Dr. Hofstadter, is mixing them up, not only fancifully in language but in real natural systems. How do we know that nature allows that privilege? After all, language paradoxes have an arbitrary, artificial tone to them." You are very happy to be able to argue, if not with Hofstadter, at least with his proponent with what seems to you irrefutable logic.

But Hofstadter's proponent is ready for you.

"Who says we can keep logic pure?" he objects. "Or haven't you heard of Gödel's theorem? I thought you read Dr. Hofstadter's book."

"I told you I didn't understand it. And it was Gödel's theorem that stopped me cold."

"It is really very simple. Logical typing was invented by two mathematicians, Bertrand Russell and Alfred Whitehead, to keep logic pure, as you say. What another mathematician, Kurt Gödel, proved is that any attempt to produce a paradox-free mathematical system is bound to fail if that system is reasonably complex. He proved this by showing that any system of reasonable richness is doomed to be incomplete. You can always find a statement within it that the system cannot prove. In fact, the system can be either complete but inconsistent or consistent but incomplete but can never be both consistent and complete. The way Gödel proved his theorem was to use the so-called impure logic of tangled hierarchies. So right out the window went a number of ideas, including the possibility of a complete and consistent mathematical system like Russell's and Whitehead's theory of logical types. Any questions?"

You do not dare to ask any further questions. Mathematics is a hornet's nest for you. The longer you linger, the more risk you take of being stung. Eagerly, you thank the gentleman and head for the nearest exit.

But, of course, I stop you just before you reach the exit. You are surprised at seeing me. "What are you doing here?" you ask.

"It's my book. I can butt in whenever I like," I tease. "Tell me. Did you buy Hofstadter's pitch about building a self-aware silicon computer?"

"Not entirely, but it seemed like an interesting idea," you reply.

"I know. The idea of the tangled hierarchy is fascinating. But did anybody explain how Hofstadter is going to generate discontinuity in the programs of a classical silicon machine that are by their very nature continuous? It's not so much that the programs feed back on each other and get so tangled that for all practical purposes you cannot follow their causal chain. It's not like that at all. There really has to be a discontinuity, a real jump out of the system, an inviolate level. In other words, the question is how can our brain, looked at as a classical system, have an inviolate level? In the philosophy of material realism on which classical systems are based, there is only one level of reality, the material level. So where is the scope for an inviolate level?"

"Don't ask me," you plead. "What are you suggesting?"

"Let me tell you a story. The Sufi master Mulla Nasruddin was found one day on his knees adding yogurt to the water in a pond. A passerby asked, 'What are you doing, Nasruddin?'

'I'm trying to make yogurt,' answered the mulla.

'But you can't make yogurt that way!'

'But suppose it takes,' the mulla said optimistically."

You chuckle. "A funny story. But stories don't prove anything," you object.

"Have you heard of Schrödinger's cat?" I fire back.

"Yes," you say, brightening up a bit.

"According to quantum mechanics, the cat is half dead and half alive after its hour is up. Now suppose a machine is set to observe if the cat is alive or dead."

"I know all about that," you cannot resist. "The machine picks up the cat's dichotomy. It's unable to align its pointer to a definite reading, dead or alive, until a conscious observer relieves it."

"Good. But now suppose we send a whole hierarchy of inanimate machines successively to observe the reading of each previous machine. Isn't it logical that all of them will acquire the quantum dichotomy of the cat's state?"

You nod your head in approval. It seems logical enough.

"So by having the cat's wave function in a quantum superposition, we have in effect opened up the possibility that all material objects in the universe are susceptible to contracting the contagious quantum superposition. The quantum superposition has taken on a universality. But there is a price. Do you see?"

"No, I don't."

"The system is not closed."

"Ah."

"This openness or incompleteness is a logical necessity if you play Schrödinger's game, attributing quantum description to macro systems. Now this is a true Gödelian knot."[5]

"What are you getting at?" you ask, puzzled.

"To untie the knot we have to be able truly to jump out of the system, and that means a quantum machinery in our brain with nonlocal consciousness collapsing it. So we must have a quantum system in our heads in order to have a genuine tangled hierarchy—discontinuity, inviolate level, and all."

"Really?"

But I terminate the inquiry (discontinuously, using the privilege of the inviolate level). All things that have a beginning must end somewhere for the moment, even exciting concepts like a quantum system in our brains.

Okay, so you now know what a tangled hierarchy is, you are satisfied that it only works for a quantum system within an overall idealist framework, and you intuit that it may be the explanation of our own self-reference. Let's try it and see.

SCHRÖDINGER'S CAT REVISITED

To see how tangled hierarchy and self-reference arise in the brain-mind, let us return once more to Schrödinger's cat.

According to quantum mechanics, the state of the cat is half dead and half alive after the hour. Now we set a machine to measure if the cat is alive or dead. The machine picks up the contagious dichotomy of the cat. And if we set a whole series of insentient machines, one after the other, to measure the reading of each previous machine, the logic is inescapable that all of them will acquire the quantum dichotomy.

It is a little like the story of the islander and the missionary. The missionary is explaining how the earth is held up by gravity and so forth. But the islander confronts him, declaring: "I know who really holds up the earth. It's a turtle."

The missionary smiles condescendingly. "But then, my dear man, who is holding up the turtle?"

The islander is unperturbed. "You're not going to trick me with that one," he admonishes. "It's turtles all the way down."

The point of the von Neumann chain, of course, is that the dichotomy of the measuring apparatuses that observe Schrödinger's cat goes 'all the way down.' The system is an infinitely regressive one. It does not collapse of itself. We vainly chase the collapse in a von Neumann chain just as we chase the truth value in the liar's paradox. In both cases we end up in infinities. We have the makings of a tangled hierarchy.

To resolve the knot, we have to jump out of the system to the inviolate level. According to the idealist interpretation of quantum mechanics, the nonlocal consciousness acts as the inviolate level, since it collapses the brain-mind from outside space-time, thus terminating the von Neumann chain. There is no Gödelian knot from this perspective.

Things are different, however, from the perspective of the brain-mind. Let us make a crude model of the brain-mind's response to a stimulus. The stimulus is processed by the sensory apparatus and presented to the dual system. The state of the quantum system expands as a coherent superposition, and all the classical measuring apparatuses that couple with it also become coherent superpositions. There is no mental program, however, that chooses among the different facets of the coherent superposition; there is no program in the brain-mind that we can identify as a central processing unit. The subject is not a homunculus acting at the same level as the brain-mind's programs.

Instead, there is a discontinuity, a breakdown of causal connection within space-time in the process of selection from the possible choices in the probability pool that the quantum system gives. The choice is a discontinuous act in the transcendent domain, an act of our nonlocal consciousness. No linear, cause-effect description of it in space-time is possible. This is the 'white spot' (as in Escher's drawing *Print Gallery*) in our picture of a tangled hierarchy in the brain-mind. The result is self-reference. Consciousness collapses the total quantum state of the dual system, resulting in the primary separation of subject and object. Because of the tangled hierarchy, however, consciousness identifies itself with the "I" of the self-reference and experiences the primary awareness, *I am.*

Realize that *the self of our self-reference is due to a tangled hierarchy, but our consciousness is the consciousness of the Being that is beyond the subject-object split.* There is no other source of consciousness in the universe. *The self of self-reference and the consciousness of the original consciousness, together, make what we call self-consciousness.*

Chapter 13

THE "I" OF CONSCIOUSNESS

THE CONCLUSION of the preceding chapter bears repeating, for it provides the basis of understanding ourselves in the universe: The self of our self-reference is due to a tangled hierarchy, but our consciousness is the consciousness of the Being that is beyond the subject-object split. There is no other source of consciousness in the universe. The self of self-reference and the consciousness of the original consciousness, together, make what we call self-consciousness.

In a sense, we are rediscovering ancient truth. It is indeed wondrous that humankind implicitly has always known that self-consciousness results from a tangled hierarchy. This knowledge, inherent in many cultures, has appeared at different places and times in the archetypal picture of a snake biting its own tail (fig. 34).[1]

It is the appearance of the world of manifestation that leads us to the experience of a self or a subject that is separate from the objects of appearance. That is, subject and object manifest simultaneously in the initial collapse of the quantum state of the brain-mind. As the romantic poet John Keats intuited: "See the world if you please/As a vale for soulmaking."

Without the immanent world of manifestation, there would be no soul, no self that experiences itself as separate from the objects it perceives.

For convenience a new term can be introduced to describe this

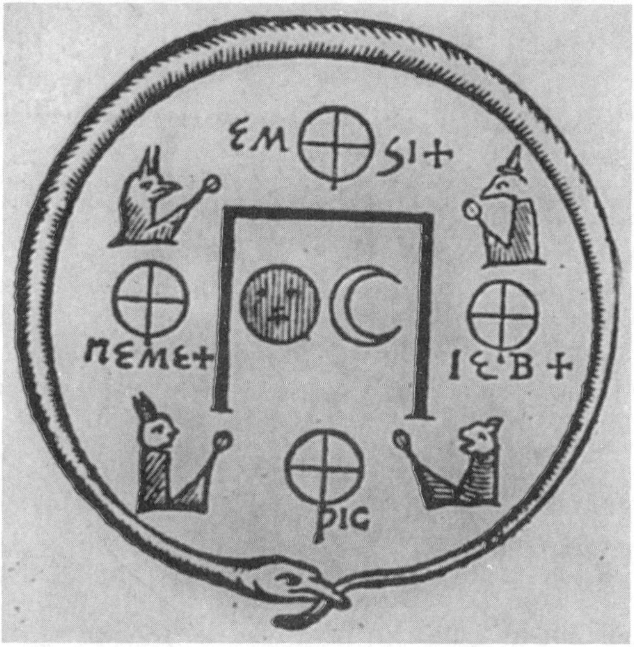

Figure 34. The Uroboros. (From Neumann, Eric, *The Origins and History of Consciousness*, translated by R. F. C. Hull. Bollinger series XLII, © 1954, 1982, renewed by Princeton University Press. Reprinted with permission of Princeton University Press.)

situation. Before collapse, the subject is not differentiated from the archetypes of objects of experience—physical or mental. Collapse brings about the subject-object division, and that leads to the primary awareness of I-am-ness that we will call the *quantum self.* (Of course, we could also say that the awareness of the quantum self brings about collapse. Remember the inherent circularity of self-reference.) Consciousness identifies with the emergent self-reference of its quantum self, in which the unity of the subject still persists. The next question is, How does our so-called separate self—our unique reference point for experience, the individual ego—arise?

THE EMERGENCE OF THE EGO

"We cannot escape the fact that the world we know is constructed in order (and thus in such a way as to be able) to see itself," says the mathematician G. Spencer Brown, "but in order to do so, evidently it must first cut itself up into at least one state which sees, and at least one other state which is seen."[2] The mechanisms for this subject-object division are the double-whammy illusions of the tangled hierarchy and of the identity of the self with the locus of our past experiences that we call the ego. How does this ego-identity arise?

I have said that the brain-mind is a dual quantum system/measuring apparatus. As such, it is unique: It is the place where the self-reference of the entire universe happens. *The universe is self-aware through us.* In us the universe cuts itself into two—into subject and object. Upon observation by the brain-mind, consciousness collapses the quantum wave function and terminates the von Neumann chain. We resolve the von Neumann chain by recognizing that consciousness collapses the wave function by acting self-referentially, not dualistically. In what way does a self-referential system differ from a simple combination of quantum object and measurement apparatus? The answer is crucial.

The brain's measurement apparatus, like all other measurement apparatuses, makes a memory of every collapse—that is, every experience that we have in response to a certain stimulus. Additionally, however, if the same or a similar stimulus is presented again, the brain's classical record replays the old memory; this replay becomes a secondary stimulus to the quantum system, which then responds. The classical system measures the new response, and on it goes in that manner. This repeated measurement interaction leads to a fundamental change in the brain-mind's quantum system; it is no longer regenerative.[3]

Each previously experienced, learned response reinforces the probability of the same response over again. The upshot is as follows: For a novel, unlearned stimulus, the behavior of the brain-mind's quantum system is like that of any other quantum system. As a stimulus is learned, however, the likelihood increases that, after the completion of a measurement, the quantum-mechanical state of the dual system will correspond to a prior memory state. In other words, learning (or prior experience) biases the brain-mind.

This explanation is, of course, a theoretical analysis within the present brain-mind model of simple behavioral conditioning. Before the response to a particular stimulus becomes conditioned, before we experience it for the umpteenth time, the probability pool from which consciousness chooses our response spans the mental states common to all people at all places at all times. With learning, conditioned responses gradually begin to gain greater weight over others. This is the developmental process of the individual mind's learned, conditioned behavior.

Once a task has been learned, then for any situation involving it, the likelihood that the corresponding memory will trigger a conditioned response approaches 100 percent. In this limit, the behavior of the dual quantum system/measuring apparatus becomes virtually classical. Here you see the brain-mind analog of Bohr's correspondence principle. In the limit of a new experience, the brain-mind's response is creative. With learning, the probability of a conditioned response is increasingly enhanced, until—in the limit of an infinitely repeated experience—the response is totally conditioned, as behaviorism posits. This is important because classical conditioning as posited by behaviorism is recovered as a special case of the more general quantum picture.

Fairly early in an individual's physical development, many learned programs accumulate and dominate the brain-mind's behavior—despite the fact that unconditioned quantum responses are available for new creative experiences (especially in response to unlearned stimuli). But when the creative potency of the quantum component is not engaged, the tangled hierarchy of the interacting components of the brain-mind, in effect, becomes a simple hierarchy of the learned, classical programs: Mental programs respond to one another in a well-defined hierarchy. At this stage, the creative uncertainty as to "who is the chooser" of a conscious experience is removed; we begin to assume a separate, individual self (ego) that chooses and that has free will.

To further explain this concept, suppose that a learned stimulus arrives at the brain-mind. In response, the quantum system and its classical measuring apparatus expand as coherent superpositions but weighted heavily in favor of the learned response. The memories of the classical computer respond also with learned programs associated with the given stimulus. After the event of collapse associated with the primary experience, a series of secondary-collapse processes takes place. The quantum system develops in relatively unambiguous states in response to the classical, learned programs, and each is amplified and collapsed. This series of processes results in secondary experiences that have a distinctive quality, such as habitual motor activity, thoughts (for example, I did this), and so on. The learned programs that contribute to the secondary events are still part of a tangled hierarchy, for by following them we find a break in their causal chain that corresponds to the role of the quantum system and its collapse by nonlocal consciousness. This discontinuity, however, is obscured and interpreted as an act of the free will of a (pseudo-) self; it is then followed by the (false) identification of the nonlocal subject with a limited individual self associated with the learned programs. This is what we call ego. Clearly, the ego is our *classical self.*

To be sure, our consciousness is ultimately unitive and is at the transcendent level, which we now recognize as the inviolate level. From inside physical space-time (from the point of view of the classical programs of our brain-mind), however, we become possessed by individual identity: ego. From inside, little able to discover our system's tangled hierarchical nature, we claim free will to mask our assumed limitedness. The limitedness arises from accepting the point of view of the learned programs causally acting on one another. In ignorance we identify with a limited version of the cosmic subject; we conclude, I am this body-mind.

As the real experiencer (the nonlocal consciousness) I operate from outside the system—transcending my brain-mind that is localized in space-time—from behind the veil of the tangled hierarchy of my brain-mind's systems. My separateness—my ego—emerges only as the apparent agency for the free will of this cosmic

"I," obscuring the discontinuity in space-time that the collapse of the quantum brain-mind state represents. A quote from one of Wallace Stevens's poems is relevant to the question of one's separateness:

> *They said, "You have a blue guitar*
> *You don't play things as they are."*
> *The man said, "Things as they are*
> *Are changed upon the blue guitar."*[4]

Things as they are (such as the pure, undivided cosmic consciousness) are manifest as the separate, individualized ego; they are changed by the blue guitar of the simple hierarchy of the learned programs of the individual brain-mind.

The separate self, however, is only a secondary identity for consciousness because the nonlocal, creative potency of consciousness and the versatility of the quantum mind never completely disappear. They remain present in the primary quantum modality of the self.

Classical and Quantum Selves

The psychologist Fred Attneave defines the ego as follows: ". . . stored information about past states of consciousness may be recalled into consciousness. Thus it becomes possible for consciousness to see its own reflection in the mirror of memory—though always (violating the metaphor a little) with a time lag. It is in these terms, I believe, that the [ego] is to be defined."[5]

Note, in particular, the time lag referred to by Attneave: It is the reaction time between the collapse of a space-time event (the onset of the quantum modality) and the verbally reported secondary classical mode, or introspection-based experience of the ego. There is impressive evidence supporting the concept of this introspection time.

Neurophysiologist Benjamin Libet, neurosurgeon Bertram Feinstein, and their collaborators have discovered the intriguing phenomenon of introspection time in patients undergoing brain surgery at Mount Zion Hospital in San Francisco.[6] (Brain surgery patients can remain awake during surgery because there is no pain involved.) Libet and Feinstein measured the time it takes for a touch stimulus on a patient's skin, traveling as a spiked electrical activity along the neuronal pathway, to reach his brain. It was about $^1/_{100}$th of a second. What Libet and Feinstein found is that the patient did not verbally report being consciously aware of the stimulus for close to half a second. In contrast, a behavioral response by such subjects (such as pushing a button or saying the word "go") takes only $^1/_{10}$th to $^2/_{10}$ths of a second.[7]

Libet's experiments support the concept that the normal classical ego-self arises from processes of secondary awareness of a conscious experience. The nearly half-second between the behavioral response and the verbal report is the time taken for processing secondary awareness; it is the (subjective) reaction time taken for the I-am-this type of introspection. Our preoccupation with the secondary processes (indicated by the time lag) makes it difficult to be aware of our quantum self and to experience the pure mental states that are accessible at the quantum level of our operation. Many meditation practices are intended to eliminate the time lag and to put us directly in touch with those pure mental states in their suchness (*tathata* in Sanskrit). Evidence (albeit tentative) shows that meditation reduces the time lag between the primary and the secondary processes.[8]

Circumstantial evidence also shows that exalted experiences occur when this time lag is reduced. George Leonard has reported the exalted experiences of athletes.[9] For example, when a baseball outfielder makes an outstanding catch, the exaltation may not be the result of the success (as is usually assumed) but the result of the reduced reaction time (which makes the catch easy for him) that enables him to glimpse his quantum self. The outstanding catch and the exaltation arise simultaneously—each in effect causing the other. Maslow's data on peak experiences—direct transcendental experiences of the self as rooted in the unity and harmony of a cosmic Being (for example, the creative ah-ha experience)—can also be explained in terms of reduced reaction time and the quantum self of the experiencer.[10]

The time lag of secondary introspection allows our ego-experience of consciousness to feel continuous. Our so-called stream of consciousness is the result of mindless introspective chatter. (What a price to pay for the accumulation of experience!) Consciousness divides itself into subject-object via a collapse of the brain-mind's quantum wave function. The collapse is an event of discontinuity in space and time, but we lopsidedly experience the subject-object division in the continuous, classical ego modality. We are hardly aware of the immediacy of experience available in the quantum mode, which T. S. Eliot has recognized as the "still point," referred to in the following excerpt from one of his poems:

Neither from nor towards; at the still point, there the dance is,
But neither arrest nor movement. And do not call it fixity,
Where past and future are gathered. . . .
. . . Except for the point, the still point,
There would be no dance, and there is only the dance.[11]

Maya is now explained.[12] The immanent world is not *maya*; not even the ego is *maya*. The real *maya* is the separateness. Feeling and thinking that we are *really* separate from the whole is the illusion. We have attained the final objective of quantum functionalism—finding an explanation of our separate self. With its classical learned programs forming an apparent simple hierarchy, consciousness acquires ego (an I-am-this-ness) that identifies with the learned programs and the individual experiences of a particular brain-mind. Such a separate self has aspects of an emergent phenomenon, as Sperry suspected. It emerges out of the introspective interaction of our learned programs that result from our experience in the world, but there is a twist. The separate self has no free will apart from that of the quantum self and, ultimately, that of the unitive consciousness.

I hope you now see the essence of quantum functionalism. Whereas conventional theories of brain-mind avoid the concept of consciousness as an embarrassment, quantum functionalism begins with consciousness; yet it recovers the behaviorist description of the actions of the brain-mind as a limiting case and even agrees with the materialists that the ego's free will is a sham. The new theory is far more versatile as an aid to understanding the brain-mind, however, because it also acknowledges the quantum modality of the self.

Materialist psychologists believe only in the ego, if even that. Many of them would say that there is no quantum self. Imagine, however, that there were a potion that could sever the quantum self. What would life be like? The following parable plays with that question.

THE LOVE OF A CLASSICAL MECHANIC: A PARABLE

Once upon a time, there was a woman who believed in classical mechanics and classical logic. All the talk of many of her friends, sometimes even of her husband, about idealist philosophy, mysticism, and such made her uneasy and uncomfortable.

In her relationships with people, she could not understand what they wanted. She always treated her parents well, but they wanted her to share herself with them. She didn't know what they meant. She liked sex with her husband, but he talked too much about trust and love. Those were just words. What was the use of words like that? Sometimes, lying awake after having sex with the man who was her husband, she felt inundated by feelings of emotional tenderness. She imagined that they were the same kind that made her parents sometimes look at her in misty-eyed silence. And she hated the mushiness of it.

She could not understand why some of her friends looked for meaning in their lives. Some of them incessantly talked of love and of aesthetics. She had to control her laughter for fear of offending them, but she knew that they were being naive. There was, she thought, no love apart from sex. Yet sometimes when she was gazing at the ocean unaware, she would melt into a feeling of unity with the vastness of the ocean. Then she would lose a moment or two of her existence and would be immersed in love. She hated and feared those moments.

She had tried to communicate her uneasiness a couple of times, but her confidantes had talked reassuringly of her inner quantum self, beyond her ordinary ego. She would never believe in anything that elusive. Even if she did have some sort of inner self, she wanted no part of it. Then one day she heard about a newly discovered potion that would disconnect one from the quantum self. She sought out the person who had discovered the potion.

"Will your potion enable me to enjoy sex without feeling mushy about love?"

"Yes," said the man who had the potion.

"I can't bear the insecurity of trusting people. I'd rather count on trade-offs and back-ups. Will your potion enable me to live life without having to trust people?"

"Yes," said the man who had the potion.

"If I take your potion, will I be able to relax in the beauty of the ocean without having to cope with those feelings of so-called universal love?"

"Always," said the man who had the potion.

"Then your potion is for me," she said, eagerly quaffing it.

Time passed. Her husband began to sense a change in her. Her behavior was about the same, but he could not, as he said, feel her vibes as he used to. Then one day she told him that she had taken a potion to disconnect her quantum self. Immediately, he sought out the man who had given his wife the potion. He wanted his wife to regain her quantum creativity.

The man who gave his wife the potion listened to him for a while and then said: "Let me tell you a story. There was a man once who had a pain in one of his legs that he couldn't bear. The doctors couldn't find a cure. They finally decided to amputate. After long hours under anesthesia, the patient woke to see his doctor looking at him quizzically. Still not feeling too well, he asked the doctor, 'Well?'

" 'I have some good news and some bad news. First, the bad news. We cut off the wrong leg.' The patient stared at him uncomprehendingly, but the doctor was quick to reassure him. 'And now the good news. The bad leg is not so bad, after all. There is no need to amputate. You will be able to use it.' "

The husband looked puzzled. The man who gave his wife the potion continued. "Your wife didn't like the creative uncertainty of life that comes with the quantum self, so she relieved herself of it. She preferred to walk on one leg, as it were. That's bad news for you. But now the good news. I do have a remedy for husbands like you. I can condition her into the soulful behavior that you want from her. With my training, she will give you both tea and sympathy."

The husband was delighted. And so it was done. His wife seemed like her old self again. She occasionally whispered little words of love as she had before she took her potion. But her "soulful" husband still could not feel her vibes.

He went back to the man who had given the potion to his wife and had taught her loving behavior. "But I am not really satisfied with behavior alone. I want something ineffable—I want to feel her vibes," the husband lamented.

The man said, "There is only one thing to do. I can give you the potion and then train you as I did your wife."

Since there was no alternative, the husband agreed. And then the couple lived happily ever after. Nobody in their town had ever seen a more loving couple. They were even selected as life members of the local chapter of Walden II, the first such honor ever bestowed.[13]

Not to worry, such a potion will never be found. Yet, incessant and unnecessary behavioral, cultural, political, and social conditioning do function as the chemical potion in the parable by hobbling the potential that the quantum self offers us. So the next question is, How can we take responsibility for the emerging knowledge that we are bigger than materialism acknowledges? Where do we go from here? This is the subject of Part 4.

Chapter 14

INTEGRATING THE PSYCHOLOGIES

THE SELF (the "I") is not a thing but a relationship between conscious experience and the immediate physical environment. In a conscious experience, the world appears to be divided into subject and object(s). Upon reflection in the mirror of memory, this division produces the dominant experience of the ego.

There has been much philosophical thinking on the nature of the self (or "I"). This branch of philosophy is sometimes called phenomenology. Phenomenologists study the mind via introspection, not unlike the meditation employed by Eastern mystical philosophers and psychologists. There are also numerous Western psychological models (besides behaviorism). The psychoanalytical model proposed by Freud, for example, maintains that the self is dominated by unconscious drives.

It is interesting to examine how the model of the self that we have called quantum functionalism accounts for the varieties of "I"-experience and to compare quantum functionalism with other philosophical and psychological models. This chapter includes such a comparison, incorporating some thoughts from philosophy, psychology, and the new physics (as it regards the nature of the self and of free will).[1]

CHARACTERISTICS ASSOCIATED WITH THE EXPERIENCES OF "I"

The salient experiences of the "I" are as follows:

1. Intentionality (purposeful, directional focusing toward an object, including desire, judgment, and speculation)
2. Self-awareness (sense of self)
3. Reflectivity (awareness of being aware)
4. Ego-experience (feeling that the self is a unique entity with a certain character, personality, and contingent personal history)
5. Attention (experience of the ability of the self to direct its focus toward one object or another)
6. Transpersonal-self experiences (moments of revelation or insight, as in the creative ah-ha experience)
7. Implicit experience of the self (experiences in which there is division of the world into subject and object but no explicit experience of "I")
8. Choice and free will
9. Experiences related to the unconscious

These "I" experiences are not, of course, mutually exclusive. Quite the contrary. They are intimately connected with one another. Bearing that in mind, let us now look closer at each of these experiences.

Intentionality, Self-Awareness, and Reflectivity

The pointing toward an object that is a concomitant of most conscious experience is referred to in the philosophical literature as *intentionality.*[2] There are many modes of intentionality, such as desire, judgment, and speculation. Thus the word does not refer to intentions alone. The experience of "I" that intends is, of course, self-aware, but it is much more; it is directed and purposeful in its thoughts and feelings.

So, one of the most common experiences of the "I" is of itself as a subject with intentions toward some object. Another common experience of "I" occurs when we reflect about ourselves, when, in reflective experiences, we become aware of having been aware.[3] This, too, is a subject-object experience, with the "I" playing the role of subject and consciousness playing the role of object.

What causes the division of the world into subjects and objects? Different philosophies give different answers. The major positions, those of the materialist and the idealist, are summarized here.

To material realists, the question to be answered is, How does the subject arise from a conglomerate of material objects like neurons and gray matter? Their answer is epiphenomenalism—the subject is an emergent epiphenomenon of the brain. No one, however, has been able to show how such an emergence might occur. Artificial intelligence models (connectionism[4]) depict the brain as a parallel-processing computer network; within this basic philosophy, bottom-up theorists try to prove that subject-consciousness arises as "order within chaos," as a new emergent function.[5,6] Fundamentally, all of these models suffer from the same basic conundrum: There is no provable connection between computer states (or neuronal states) and the states of mind that we experience.

In contrast, to monistic idealists all things are in and of consciousness. Thus in this philosophy, the relevant question is, How does consciousness, which is all, split itself into a subject that experiences and objects that are experienced? Here, the quantum theory of self-consciousness is able to give prima facie proof of how such a division may arise. According to this theory, the states of the brain-mind are considered to be quantum states, which are probability-weighted, multifaceted, possibility structures. Consciousness collapses the multifaceted structure (a coherent superposition) choosing one facet but only in the presence of brain-mind awareness. (Awareness, remember, is the mind-field in which objects of experience arise.) Which comes first: awareness or choice? This is a tangled hierarchy. It is this tangled-hierarchical situation that gives rise to self-reference, to the subject-object split of the world.

Further secondary-awareness processes lead to intentionality—the tendency to identify with an object. The "I" of reflective awareness also arises out of these secondary-awareness processes. Both the primary experience and the secondary processes normally remain in what is often referred to in the psychological literature as the preconscious; this obscuration of the tangled hierarchy of the primary process is fundamental to the simple hierarchical identity with our "I."

Ego-Experience

The Polish psychologist Z. Zaborowski, who reviewed the psychological literature on self-awareness, defined self-awareness as the coding, processing, and integration of information about the self.[7] In my view, such a characterization fits more than self-awareness; it also fits what is ordinarily called the ego experience. Self-awareness is a concomitant of the ego-experience but not all of it.

The most compelling experience of "I" is as the ego—the apparent doer, coder, processor, and integrator of our programs (to use Zaborowski's computer metaphor). The ego is the image we construct of the apparent experiencer of our everyday actions, thoughts, and feelings.

The ego has been the central actor in many theories of personality. Radical behaviorism and social learning theory imply that the ego is the locus of socially conditioned behavior—the result of stimulus, response, and reinforcement.[8] In more recent behavioral literature, however, the ego is seen to be the mediator of external behavior via internal mental thoughts.[9] Thus Zaborowski's cognitive definition of self-awareness and the later behavioral definition of the ego are similar.

Even according to the behavioral-cognitive school, however, the ego's actions can be fully stated in terms of input-output statements (albeit the output depends on the internal mental states). If this is so, there is no necessity for self-consciousness to be associated with the ego. This paradox is avoided by using the qualifier 'apparent' in the definition of the ego.

In the quantum theory of self-consciousness, the collapse of the coherent superposition of the quantum states of the brain-mind creates the subject-object split of the world. With conditioning, however, certain responses gain in probability when a learned stimulus is presented to the brain-mind.[10] Consciousness identifies with the apparent processor of the learned responses, which is the ego; the identity, however, is never complete. Consciousness always leaves some room for unconditioned novelty. This makes possible what we know as free will.

Attention and Consciously Directed Actions

As the phenomenologist Edmund Husserl has noted, self-awareness, and thus the ego, is associated with the direction of conscious attention.[11] There are also instances where the attention moves spontaneously.

In cognitive experiments that involve receiving and responding to a stimulus, subjects are typically able to ring a bell before they have self-awareness of the awareness of the stimulus and before they are able to verbalize this awareness of the stimulus. This capability suggests that there are primary- and secondary-awareness experiences and that the ego is associated with the secondary experiences of self-awareness but not with the primary experience.

Husserl, in describing the inherent association of self-awareness and the ability to direct attention (of which ability we are not self-aware), has coined the phrase *pure ego* to denote a unitary self of which self-awareness and the director of attention are two aspects: two sides of the same coin. We will continue in this book to use, as we have done so far, the simple word *self* to denote the concept of the unified self.

In the cognitive functionalist/connectionist model, there is no explanation of self-awareness. Attention is assumed to be a function of the central processing unit that defines the ego.

By contrast, in the quantum theory of self-reference, the self acts in two modalities: the conditioned, classical ego-modality referring to secondary experiences that include self-awareness; and the unconditioned quantum modality that is associated with primary-awareness experiences, such as choice and direction of attention without self-awareness. The quantum model, therefore, agrees with the model of the phenomenologists.

Transpersonal-Self Experiences

In some experiences the identity of the self with the ego is considerably less than usual. An example is the creative experience in which the experiencer often describes the act as an act of God. Another example is the "peak experience" studied by the psychologist Abraham Maslow.[12] Such experiences occur with a clear discontinuity in contrast to the more ordinary ego-continuity of the stream of consciousness. These experiences will be called *transpersonal-self experiences* since the identity with the particular persona of the experiencer is not dominant.

Transpersonal-self experiences often lead to a creative extension of the self-identity defined by the ego. This has been called self-actualization by Maslow (in the work previously cited) and, in this book, an act of inner creativity. In Eastern psychology, this creative self-making is called the awakening of intelligence—*buddhi* in Sanskrit. Since the English word *intelligence* has other connotations, we will use the Sanskrit *buddhi* to mean the extended self-identity beyond ego. Although the behavioral-cognitive model does not acknowledge transpersonal experiences, the quantum theory recognizes them as direct experiences of the quantum modality of the self.

One major characteristic of transpersonal experiences is nonlocality—communication or propagation of influence without local signals. Simultaneous scientific discoveries are possible examples of such nonlocal synchronicity. Paranormal experiences, such as telepathy, provide other examples.

Implicit Experience of the Self

As the existential philosopher Jean-Paul Sartre has pointed out, much of our common experience does not include the ego-"I." Sartre gave the example of a man counting cigarettes. As the man counts, he is absorbed in this task and has no self-awareness or any other reference to his ego. Then a friend comes around and asks, What are you doing? The man replies, I am counting my cigarettes. The man has regained his self-awareness.[13] In this kind of experience, there is consciousness, and the world is implicitly divided into subject and object; there is, however, little or no secondary reverberation of the experience.

Sartre's example falls in the lowest category of what the East Indian exponent of yoga, Patanjali (around the second century A.D.), calls *samadhi.*[14] Starting with absorption in the object (the state of lowest samadhi), one begins a journey of transcending the object in higher and higher samadhis. Eventually, a state is reached when the object is seen in its identity with cosmic nonlocal consciousness.

In Eastern psychology, the subject of the cosmic-consciousness experience is referred to as the *atman.* Christianity refers to this primary universal self entity as the holy spirit. In Buddhism, it is sometimes called no-self, since it dependently co-arises with awareness (not hierarchically superior to awareness, its object). Other Buddhist philosophers have referred to the subject of pure awareness as the universal consciousness (for example, in the *Lankavatara Sutra*). As the current Dalai Lama of Tibet points out, the terminology of no-self confuses people because it makes them think of nihilism.[15] In modern psychology, Assagioli has referred to this self-less self as the transpersonal self.[16] In the absence of an unambiguous English word, we will use the Sanskrit word *atman* to denote the self of the pure-awareness experience.

In the quantum theory of the self, the atman is seen as the quantum self—the unconditioned universal subject with which consciousness identifies and that arises codependently with awareness upon the collapse of the quantum coherent superposition. The individual self-experience, or ego, arises in the mirror of memory from secondary reverberations of primary experiences. Considerable neurophysiological evidence shows that there exists a time lag between primary- and secondary-awareness experiences.

Choice and Free Will

Perhaps the most confusing of the self experiences are those that involve choice and/or free will. Any conscious experience involves an opening out to the future and, in this sense, may be considered to involve openness or possibility. The experiences of choice and free will go beyond such openness. We will distinguish between the two terms although they often are used synonymously. *Choice* applies whenever we choose between alternatives, with or without self-awareness. *Free will* applies whenever a subsequent action is undertaken out of our own causal initiative.

Traditionally, behaviorists and cognitivists would say that there is no freedom of choice or free will. If we are classical computers—parallel processing or not—neither of these concepts makes any sense. The argument is simply that there is no causal power that can be attributed to the ego, whose behavior is completely determined by the state of its hardware and by its inputs from the environment.

Spiritual and transpersonal psychologies would agree with the behavioral assessment that the ego does not have free will, but they would insist that there is real free will. It is the free will of the atman—the consciousness that exists before any kind of reflective, individual-self experience. If the ego does not have free will, how do we in our ego go beyond ego, which is the objective of spiritual traditions? The answer that the ego is an illusion does not seem satisfactory.

With the help of the quantum theory of consciousness, we now can resolve the conceptual quandary about free will. In the quantum theory, choice defines the primary self—the atman. I choose, therefore (tangled hierarchically) I am. However, with conditioning the choice is no longer completely free but biased in favor of conditioned responses. The question is, How far does the conditioning extend?

Obviously, at the primary-process level there is no conditioning; consequently, there is unrestricted freedom of choice. At the secondary level we have conditioned responses in the form of thoughts and feelings, but do we have to act on them? Our free will at the secondary level consists of the capacity to say no to learned conditioned responses.

Notice that we are led to using the two words *choice* and *free will* somewhat differently, and this is good. Current neurophysiological experiments show that there is virtue in not using the phrase *free will* for such experiences as using one's free will to raise one's arm. Recent experiments by Benjamin Libet clearly indicate that even before a person experiences awareness of his or her action (which is necessary to free will), there is an evoked potential that signals an objective observer that the person is going to will to raise his or her arm. In view of this, how can one say that free will of this kind is free? But Libet's experiments also reveal that a person retains his or her free will to say no to raising an arm, even after the evoked potential signals otherwise.[17]

Clarifying the meaning of free will in this way can help us see the benefits of meditation—concentrating attention in the field of awareness on either a particular mind-object or on the entire field. Meditation allows us to become witnesses to the mental phenomena that arise in awareness, to the conditioned-response parade of thoughts and feelings. It creates a gap between the arousal of mental responses and the urge physically to act on them and thus enhances our capacity of free will to say no to conditioned acts. It is easy to see the value of such enhancement for changing destructive habitual behavior.

Experiences Related to the Unconscious

Some experiences are related to what is unconscious in us—to processes for which there is consciousness but no awareness. In quantum theory, these are situations in which the quantum state does not collapse but goes on developing in time according to the dynamics of the situation. The unconscious dynamics, however, may play a significant role in later conscious events. This aspect allows us to verify the effects of quantum interference in experiments of unconscious perception.[18]

In psychoanalytical thinking, some of the ego-self experiences are repressed in what Freud called the id and Jung, the shadow. The remaining conscious experiences then define the persona—the image that one projects for people to see, the image of who a person thinks he or she is. I shall refer to the repressed part of the ego-self simply as the personal unconscious. Some of our ego experiences become distorted by the influence from the personal unconscious, and this unconscious influence gives rise to the psychopathologies—such as neurosis—that psychoanalysis tries to address.

How does the personal unconscious arise according to the quantum theory? It arises as follows: The subject is conditioned to avoid certain mental states; consequently, the probability becomes overwhelming that these states are never collapsed from coherent superpositions that include them. Such coherent superpositions, however, may dynamically influence without apparent external cause the collapse of subsequent states. Not knowing the cause of behavior may lead to neurosis-generating anxiety. Eventually, the subject may imagine causes and proceed to eliminate them through such neurotic behavior as compulsive hand washing.

Similarly, Jung suggested that many of our transpersonal experiences are influenced by repressed archetypal themes of a collective unconscious—universal states that we usually do not experience. These repressed themes may also lead to pathologies.

In quantum theory, the contingent human form is subject to conditioning that suppresses certain mental states from manifestation in the world. For example, a male body would tend to suppress those mental states that pertain to explicitly female experience. This is the origin of the Jungian anima archetype. This suppression of the anima adversely limits male behavior. (Similarly, the animus archetype in females is suppressed, divorcing women from the male experience.)

When we dream or when we are under hypnosis, the self becomes primarily a witness and enters a state of relative absence of secondary-awareness events. In such a state, the normal inhibitions against collapsing repressed mental states are weakened. Thus both dreams and hypnosis are useful for bringing the unconscious to conscious awareness.

Similarly, in near-death experiences the immediacy of death releases much repressed unconscious conditioning, both collective and personal. As a result, many patients come out of near-death experiences full of joy and peace.

In attaining freedom in our actions, it is important to avoid being dominated either by our ego/persona conditioning or by our tyrannical, internal, repressed unconscious coherent superpositions.

THE SPECTRUM OF SELF-CONSCIOUSNESS

By surveying the characteristics of conscious experiences as described by phenomenology, psychology, cognitive science, and quantum theory, we can arrive at an important summary of how the self manifests in us—a summary that is of the spectrum of self-consciousness (see also Wilber[19]). Of all these theoretical models, however, only one—the quantum theory of consciousness—has the breadth to encompass the entire spectrum; therefore, the idealist quantum view of consciousness will be adopted from the outset in this summary.

In monistic idealism, consciousness is one—one without a second, said Shankara.[20] The spectrum of self-consciousness consists of stations with which the one consciousness identifies itself at various stages of human development. The entire spectrum is surrounded at the lower end by the personal unconscious and at the upper by the collective unconscious. All the stages, however, are in consciousness.

This schema is conceived in developmental, not in hierarchical, terms. The higher we develop, the more ego-less we become, until at the highest level there is no discernible identity with the ego at all. Thus a profound humility characterizes the levels of being beyond ego.

The Ego Level

At this level, the human being identifies with a psychosocially conditioned and learned set of contexts on which to operate. These contexts give the human person a character. Depending on how absolute this ego-identity is, the person at this level tends to be solipsistic. The contexts within which this person operates tend to take on an aura of infallibility, and all other contexts are judged against the criteria of these personal contexts. The person believes, Only I and my extensions (my family, my culture, my country, and so forth) have primary validity. All others are contingent.

Within the basic ego level, we can identify two bands. The first one, the pathological band, is closer to the personal unconscious. It is strongly affected by internal stimuli (uncollapsed coherent superpositions) from the unconscious. People in whom the self identifies with this band are often disturbed by the strivings and motivations of the unconscious. Their ego is divided into a self-image and a shadow image—the first propagated and the second suppressed.

The second band, the psychosocial, is where most of us live except for an occasional excursion into lower and upper (in the developmental sense) bands of identity. In upper excursions, for example, we may be able to say no to a conditioned habitual response, thus exercising our free will; or we may delve into creative activities in the world; or we may unselfishly love somebody. The usual motivations for action at this level, however, are directed by a personal agenda that serves the perpetuation and strengthening of the character-image identity in its striving for fame, power, sex, and so forth.

The Buddhi Level

This level is characterized by a less restricted identity for the self—one that explores the entire human potential. The personal motif of living at the ego level is replaced by one of inner creativity, self-exploration, and actualization.

Within this level, we can identify several bands. The bands, however, are not hierarchical, nor are they necessarily experienced in any chronological order. Some of them may even be bypassed.

The first, closer to the ego level, will be called the psychic/mystical band. People who identify their self with this band have nonlocal psychic and mystical experiences that enlarge their vision of the world and their role in it. The themes of the collective unconscious often surface in dreams, creative experiences, and the understanding of myths, which provide additional motivation for freedom and integration of the self. Yet at this level of self-identity, people are still too motivated by personal desires to shift decisively to a truly fluid identity.

The second band is transpersonal. There is now a certain ability and tendency to witness internal processes without necessarily externalizing them. One's psychosocial contexts of living are no longer absolute. Otherness is discovered, and some of the joys of this discovery (such as the joy of service) enhance motivation.

The third band, the spiritual, is an identity that few people on earth have been known to display. Life is lived primarily in an easy-without-effort (*sahaj*, in Sanskrit) samadhi. The self is more or less integrated; the themes of the collective unconscious are much explored; and actions are appropriate to events. Because of the rarity today of individuals whose identities reside in this band, we have very little scientific data about it. There are, of course, many historical cases of this identity in the mystical and religious literature of the world.

The highest level is the atman, the level of the self (or no-self) attainable only in samadhi.

Note that the spiritual psychologies of India and Tibet refer to seven bands of self-identity (one extra band at the ego level). The origin of this system lies in the Indian idea of three kinds of drives, the three *gunas*: *tamas*, or inertia; *rajas*, or libido; and *sattwa*, or creativity.[21] The Indian psychologists posit three ego bands—perhaps one for each type of drive dominance, but since it is acknowledged that all people have some of each of the gunas, this kind of classification seems somewhat redundant.

The question may be raised, How does a shift of the self-identity occur? There is a Zen story that addresses the question: "The student Doko came to the Zen master and said: 'I am seeking the truth. In what state of the self should I train myself, so as to find it?' Said the master: 'There is no self, so you cannot put it in any state. There is no truth, so you cannot train yourself for it.' "

In other words, there is no method, no training for the shift in self-identity. That is why we call the process inner creativity. The process is one of the breakdown of the boundary that is determined by one set of contexts for living to allow an expanded set of contexts.[22] We will go into further details of this process in Part 4.

Note that the integration achieved here of the theories of personality and the self within the context of the quantum theory of consciousness should lead as well to integration of the various schools of psychology—psychoanalytic, behavioral, humanistic/transpersonal, and cognitive. Although we have shown that the model based on cognitive science and artificial intelligence is not adequate for the complete description of the human person, the model still serves as a useful simulation of most of the ego-related aspects of the self.

PART 4

THE RE-ENCHANTMENT OF THE PERSON

The first draft of this book was written in the summer of 1982, but I knew there were deep inconsistencies in the material. The inconsistencies came from a very subtle attachment to one of the fundamental tenets of realistic philosophy—that consciousness has to be an epiphenomenon of matter. The biologist Roger Sperry spoke of emergent consciousness—causally potent consciousness emerging from matter, the brain. How could this happen? There is a stubborn circularity in the argument that something made from matter can act on it with causal novelty. I could see the connection with the paradoxes of quantum physics: How could we, our observations, have an effect on the behavior of objects without postulating a dualistic consciousness? I also knew that the idea of a dualistic consciousness, separate from matter, creates its own paradoxes.

Help arrived from an unexpected direction. As a scientist, I have always believed in a total approach to a problem. Since by now my research was clearly an exploration of the nature of consciousness itself, I felt that I must delve into empirical and theoretical studies of consciousness as well. This involved psychology, but the conventional psychological models—because of their roots in material realism—shy away from conscious experiences that challenge that worldview. Other, less conventional psychologies, however, such as the work of Carl G. Jung and Abraham Maslow, presupposed a different set of assumptions. Their views are more in resonance with the philosophy of the world's mystics—a philosophy that is based on spiritually seeing through the veil that creates duality. To remove the veil, the mystics prescribe being attentive to the field of awareness (such attentiveness is sometimes called meditation).

Eventually, after years of effort, a combination of meditation, reading of mystical philosophies, a lot of discussions, and just hard thinking began to break through the veil that separated me from the resolution to the paradoxes that I was seeking. The fundamental tenet of material realism—that everything is made up of matter—had to be given up, and this without bringing in dualism. I still remember the day when the final breakthrough occurred. We were visiting our friend Frederica in Ventura, California.

Earlier in the day, Maggie and I had gone with a mystic friend, Joel Morwood, to hear Krishnamurti speak in nearby Ojai. Even at age eighty-nine, Krishnamurti handled a heckler very deftly. Then in a dialog with the audience he elaborated what has been the essence of his teaching—in order to change, one must be aware now, *not deciding to change later or to think about it. Only radical awareness leads to transformation that awakens radical intelligence. When somebody asked if radical awareness comes to us ordinary beings, Krishnamurti answered gravely: "It must come."*

Later in the evening, Joel and I got into a conversation about Reality. I was giving him an earful of my ideas about consciousness, arrived at from quantum theory, in terms of the theory of quantum measurement. Joel listened with attention. "So, what's next?" he asked.

"Well, I'm not sure I understand how consciousness is manifest in the brain-mind," I said, confessing my struggle with the idea that somehow consciousness must be an epiphenomenon of brain processes. "I think I understand consciousness, but . . ."

"Can consciousness be understood?" Joel interrupted me.

"It certainly can. I told you about how our conscious observation, consciousness, collapses the quantum wave . . ." I was ready to repeat the whole theory.

But Joel stopped me. "So is the brain of the observer prior to consciousness, or is consciousness prior to the brain?"

I saw a trap in his question. "I am talking about consciousness as the subject of our experiences."

"Consciousness is prior to experiences. It is without an object and without a subject."

"Sure, that's vintage mysticism, but in my language you are talking about some nonlocal aspect of consciousness."

But Joel was not distracted by my terminology. "You're wearing scientific blinders that keep you from understanding. Underneath, you have a belief that consciousness can be understood by science, that consciousness emerges in the brain, that it is an epiphenomenon. Comprehend what the mystics are saying. Consciousness is prior and unconditioned. It is all there is. There is nothing but God."

That last sentence did something to me that is impossible to describe in language. The best I can say is that it caused an abrupt flip of perspective—a veil lifted. Here was the answer I had been looking for and yet had known all along.

When everybody else had gone to bed, leaving me to my contemplation, I went outside. The night air was cool, but I did not care. The sky was so hazy that I could barely see a few stars. But in my imagination, the sky became the radiant one of my childhood, and suddenly I could see the Milky Way. A poet of my native India fancied that the Milky Way marks the boundary of heaven and earth. In quantum nonlocality, transcendent heaven—the kingdom of God—is everywhere. "But man doesn't see it," lamented Jesus.

We do not see it because we are so enamored of experience, of our melodramas, of our attempts to predict and control, to understand and manipulate everything rationally. In our efforts, we miss the simple thing—the simple truth that it is all God, which is the mystic's way of saying that it is all consciousness. Physics explains phenomena, but consciousness is not a phenomenon; instead, all else are phenomena in consciousness. I had vainly been seeking a description of consciousness within science; instead, what I and others have to look for is a description of science within consciousness. We must develop a science compatible with consciousness, our primary experience. To discover truth, I would have to take a quantum leap beyond conventional physics; I would have to formulate a physics based on consciousness as the building block of everything. It was a difficult task, but I had just had a glimpse of the answer. So it was also simple—an easy-without-effort change of perspective. Krishnamurti's words reverberated encouragingly in *my ears: It must come. I shivered a little and the Milky Way of my imagination slowly faded away.*

The mystical truth that there is nothing but consciousness must be experienced in order to be truly understood, just as a banana, in the sensory domain, must be seen and tasted before a person really knows what a banana is. Idealist science has the potential to restore consciousness to the fragmented, Guernica-like creature that haunts each one of us. But the fragmentation of the self has its origin not only in the incomplete worldview of material realism but also in the nature of the ego-identity. If we, in our separate, fragmented egos, want to be whole again, we not only have to understand the situation intellectually, we must also delve into our inner reaches to experience the whole.

In the most celebrated of the Biblical myths, Adam and Eve live an enchanted life of wholeness in the Garden of Eden. After eating the fruit of knowledge, they are expelled from that enchantment. The meaning of the myth is clear: The price of experience in the world is a loss of enchantment and wholeness.

How can we re-enter that enchanted state of wholeness? I am speaking not of a regression to childhood or to some golden age, nor am I speaking of salvation in eternal life after death. No, the question is, how can we transcend the ego level, the level of fragmented being? How can we achieve freedom but at the same time live in the world of experience?

To answer that question, we shall in this section discuss in the context of idealist science what is conventionally called the spiritual journey. Traditionally, spiritual journeys have been designed by professional religious teachers—priests, rabbis, gurus, and others. As we shall see, the quantum scientist may add some relevant suggestions. I propose that science and religion in the future perform complementary functions—science laying the groundwork in an objective fashion for what needs to be done to regain enchantment, and religion guiding people through the process of doing it.

Chapter 15

WAR AND PEACE

IN CLIFFORD SIMAK'S Hugo Award-winning science fiction novel *Way Station* the ruling council of our galaxy worries whether earthlings will ever forget their warring ways and become civilized, learning to settle their conflicts without violence. In the novel a mystical object, a talisman, eventually effects the transformation needed for earthlings to join the civilized galaxy.

Warfare is as old as human society. Our conditioning, both biological and environmental, is such that conflicts naturally arise. For thousands of years we have employed violence to settle these conflicts, however temporarily. Now with the destructive power of atomic weapons such wars have become increasingly risky for our future on earth—not only for our lives but also for our global environment. What can we do to reduce the risks? What mystical talisman can transform our warring nations into a network of cooperative communities committed to settling conflicts through peaceful and globally responsive means?

Current social paradigms for peace are essentially reactive in that they address particular situations in which conflict has arisen or is impending. Thus the salient concerns are national security, arms control, and situational conflict resolution; all of these are reactive, situational provisions for peace. We have tried in this way to ensure peace for thousands of years, and it has not worked yet.

The situational approach to peace is locked into the materialistic and dualistic worldviews that have long dominated our view of ourselves. Today, with our image of ourselves increasingly guided by scientific realism, that view has become tunnel vision. Sociobiology (the contemporary version of social Darwinism) depicts us as selfish-gene machines—separate entities competing with one another for survival.[1] In this view, our destinies and behaviors are controlled by deterministic laws of physics and genetics and by environmental conditioning. Sociobiology is an inherently cynical amalgam of ideas from classical physics, Darwinian evolution theory, molecular biology, and behavioral psychology.

The sociobiological view of humankind is antithetical to peace in a fundamental sense. Peace as universal brotherhood and sisterhood among people, peace as cooperation springing from the heart, peace as altruism and compassion for other humans irrespective of race, color, and creed finds no room in sociobiology. In this view, the best we can hope for is situational ethics, pragmatic and legalistic containment of violence, and temporary truces in our competitive and conflicting winner/survival agendas.

In the idealist paradigm proposed in this book, we begin not with such questions as, Why is there so much conflict in the world, Why can't the Middle Eastern people learn to live with one another, Why do Hindus and Moslems constantly fight to gain the upper hand, and Why do Western nations sell lethal arms to developing countries? Instead, we ask, What creates the movement of consciousness that produces all these worldwide conflicts? Are there any compensating movements in consciousness? In other words, we seek a proactive, fundamental treatment of peace that includes all the pieces of the whole. Individually, we begin to take responsibility for these larger movements of consciousness. We are the world, so we begin to take responsibility for the world. The first step toward embracing this responsibility is to understand, intellectually at first, where other people stand relative to us as individuals. In this regard, major liberating movements in consciousness are indeed beginning to compensate (at least partially) for the old, futile movements toward violence.

UNITY IN DIVERSITY

The ideas developed in this book suggest an inner unity of human consciousness that extends beyond the diversity of individually evolved forms. The current belief in many disciplines seems to be that violence is inherent and, therefore, inevitable. If the new view is correct, however, then our separateness—the major source of the selfishness and callousness that leads to violence—is an illusion. Beyond this illusion, the separateness that is in appearance only, stands the unitive reality of inseparability.

To deal with the implication of Aspect's experiment, which establishes our inseparability beyond any reasonable doubt, the pragmatic scientist employs instrumentalism—the idea that science does not deal with reality but is only an instrument for guiding technology. But instrumentalism is insupportable. It reminds me of the student who, during an experiment with frogs and conditioning, had taught the frog to jump at his command, Frog, jump. Then he cut off one of the legs of the frog and gave his command, Frog, jump! The frog jumped and he noted with satisfaction in his lab book: Conditioning persists even when you chop off one leg. He repeated the experiment, chopping off two legs, and then three, and both times the frog jumped at his command. Finally, he chopped off the fourth leg of the frog and gave the command, Frog, jump. This time the frog did not jump. After a moment's thought, the student wrote: After losing all four legs, the frog loses its hearing.

The idea of underlying unity per se is not new; it forms the basic message of most of the world religions. Religious teachings, however, insofar as they emphasize personal salvation of some kind as the objective of self-discovery, tend to be world-negating. In contrast, when the philosophy of monistic idealism is reviewed with the new scientific attitude that has been described in this book, we get a perspective that embraces unity within the world of diversity. The new worldview affirms the world while holding up the possibility of a more mature world.

The worldview of monistic idealism and idealist science makes it clear that all manifest forms together represent only one of the many possibilities of the unitive wave that lies behind the form (of particles). The idea that unity is beyond form also implies that all the permitted diversities of form have relative value but no absolute inherent value. (This is similar to the Buddhist idea that no thing of the world has any inherent self-nature.)

When we look at the manifest world, especially the world of humans, in this way, we can easily see the wisdom in respecting and valuing the diversity of human expressions—a perspective toward cultural groups that many anthropologists have favored recently.[2] The diversity of cultures reveals human possibilities in ways that living only within the conditioning of any particular culture could never do. Every culture mirrors one image, although not a complete image, of the One. By looking at the images in different mirrors we can better understand the meaning and wonder of being human.

Thus the most modern trend of cultural anthropology is to move away from the one-language type of thinking that holds one expression, one culture, one interpretation to be the goal of human civilization (and of anthropology). The emerging direction is toward a polythematic expansion that recognizes the value of diversity for showing multitudinal dimensions of consciousness.[3] This movement from one language to multitudinal themes is paving a clear path away from the competitive war paradigm of material realism toward the cooperative peace paradigm that idealist science promises. Also important in developing an effective peace paradigm is the movement away from linear hierarchies.

FROM SIMPLE TO TANGLED HIERARCHY

If we could single out one historical concept that has propelled humans and their societies toward much violence and warfare, it is the concept of hierarchy. As the human race moved from hunting and gathering into agriculture, various hierarchies—monarchy, religious hierarchy, patriarchy, and so forth—proliferated and began to dominate human culture.

In the twentieth century, however, many social changes have involved an intuition that hierarchies are not essential, are not indispensable and universal, and at best are of only limited use. In particular, we have seen artificial hierarchies based on race and gender begin to crumble around the world.[4]

Similarly, there is a growing acceptance of the idea that the breakdown of communism in Eastern Europe and the Soviet Union, which christened the nineties, reflects not who won the arms race but what is a better system—democracy or the rigid hierarchical dictatorship of one party.

I suspect that such social revolts against hierarchies are intimately connected with the revolt in modern science against the materialist worldview. What does the new idealist science have to say about hierarchies? Often what we think of as a simple hierarchy looks simple because we are not aware of the total picture. When we do see it, as in the case of the von Neumann chain, we find that the hierarchy is a tangled hierarchy.

In discussing the important surprise element in the new model of the self based on quantum theory (chapter 12), we traced the origin of the division of reality (subject/observer and object/world) to the concept of a tangled hierarchy of interacting systems. This functional division, however, does not completely explain our sense of separateness, for the unity of the observer and the diversity of the world are complementary aspects of reality.

Our apparent separateness results from the camouflage called simple hierarchy that conceals the true mechanism of our self-reference, which is tangled hierarchy. Once this separateness arises and obscures the unity, however, it defines our perspective—thereby perpetuating itself. We become solipsistic, a collection of individual island universes with little or no awareness of our common bedrock, and we define our world in terms of our individual, separate selves: our families, our cultures, our countries. Did you notice how television programs and Hollywood movies during the eighties were narrowly defined in terms of solipsistic personal values and reflected the reign of the "me" generation?

So in this country and in the world, we have seen movements of consciousness toward women's liberation and racial equality give expression to tangled hierarchy and unity in diversity. We have also seen a contrary movement of consciousness toward the simple hierarchy of the me generation. This has been the pattern throughout history. We are like the monkey on a pole: We climb up 2 feet and then slide down 1.999 feet.

Movement away from the me generation is now under way. An idealist science has been developed, and this, too, is a movement of consciousness. So far in human history, these movements of consciousness have been largely unconscious oscillations between opposite and improperly understood polarities. Idealist science encompasses both tendencies—the solipsistic one of simple hierarchy and the one of tangled hierarchy that gives us unity in diversity—and by doing so frees us to act, each of us individually, in new creative ways.

WHERE DO I START?

The Bhagavad Gita is one of the great idealist treatises; it explores most wonderfully and comprehensively the spiritual paths for individual self-development beyond ego. Surprisingly, the book opens on a battlefield as opposing factions face each other prepared for war. Arjuna, the leader of the faction that is trying to re-establish justice, is demoralized by the prospect of killing so many people—including many relatives and friends he loves and esteems. He does not want to fight. Krishna, the teacher, is encouraging Arjuna to fight.

What kind of spiritual book promotes war instead of peace? many people ask. The answer has multiple levels of revelation.

At one level, the war in the Bhagavad Gita is not an outer war at all, but an inner battle. The conflict is in the heart of every spiritual aspirant; it is basic to all who are committed to full adult development. The predicament of Arjuna is that he is faced with killing his own kin. Is that not the case with people who aim to fulfill their human potential? One has to leave the ego-identity behind to move on, but one faces a great amount of inertia obstructing that very movement.

At a deeper level, Arjuna has a conflict with his own value system—his way of life. He is a warrior, fighting is his duty. And yet he also knows the value of love and respect and loyalty to people from whom and with whom he has learned the game of life. How can he kill those very people in battle? The situation is what Thomas Kuhn would call full of anomaly. The old paradigm is showing signs of failure and must yield to the new. So, Krishna is challenging Arjuna: Change your paradigm; you must arrive creatively at a new understanding so that you can fight without the conflict that is paralyzing you.

Is this not the case when we become entrenched in an ego-level value system that often presents conflicting demands? How does one handle crises created by anomalies, by conflicting values? We must understand that crisis is at once danger and opportunity—opportunity for creative inner transformation.

At another level, suppose there is a real war and you are fighting in it. The Bhagavad Gita gives you instructions about how to fight a war within your *dharma*, your understanding of personal, moral, and social justice. The point is, there are wars, and we are in them. Many of us have been assaulted by the questions and confusions that explode in wars around us. Remember, we are the world; true pacifism is compromised until the entire movement of consciousness is directed toward peace. So we do the best we can to serve in appropriate roles when there is an actual war.

Drawing on the wisdom of the Bhagavad Gita, interpreted for modern times, we will set out an individual manifesto for spiritual investigation toward peace—personal and global. Peace, we learn, begins with recognizing that there is conflict, both inner and outer. We will never find peace if we avoid or deny that this is so; we will never find love if we suppress the fact of hate.

Similarly, our search for joy begins with the acknowledgment that there is sorrow. (Religions start at this realization and offer ways to arrive at an uncompromised happiness that we call joy.) Our search for creative wisdom begins with the realization that in spite of all our accumulated knowledge, we do not know the answer to the particular question we are investigating; and so forth. Chapter 1 of the Bhagavad Gita is the initiation of acknowledgment of our ego-level tendencies coming from past conditioning. Similarly, we must recognize the tendency toward solipsism at both the personal and social level. Then, something can be done.

One may protest, is this not just another call to change yourself and thus change the world? Mystics and religions have preached this through the ages, but their teachings have not eliminated violence. There are several responses to this. The first I will express with a question: Have you ever considered what the world would be like if a substantial number of people through the ages had *not* taken the path of transformation? Another answer is this: I think that the mystics' call in the past has been heeded by so few largely because communication was so piecemeal. There were always barbarians (outsiders) shattering cultures before they could learn from them the advantages of peace through individual transformation. But there is no such thing as an "outside" in today's world. Communication technology has gathered us together in a global communications network.

Most importantly, this is the first time in history that we can approach inner personal growth not simply in obedience to religious authority or because we are in flight from suffering but because a coherent, growing body of knowledge and data supports such a direction of growth. In the new science, which infuses a new worldview, we draw upon science and religion and ask practitioners of both to come together as co-investigators and co-developers of a new order.

Chapter 16

OUTER AND INNER CREATIVITY

IN THE NEW, integrated psychology of the self the twin contributing factors in human development, nature and nurture, find an important third leg: creativity.[1] In psychological terms, *nature* refers to unconscious instincts that drive us—drives that Freud called libido[2]; *nurture* refers to environmental conditioning, much of which is also unconscious. Creativity, in this context, can be thought of as a drive from the collective unconscious.

In the Eastern idealist psychology of the Bhagavad Gita, there is reference to the three gunas (akin to the three drives mentioned above). The drive of past conditioning is referred to as *tamas*; this is inertia or nurture. The drive of libido is referred to as *rajas*; this is nature. The third drive is called *sattwa*: creativity.

Creativity is the creation of something new in an entirely new context. Newness of the context is the key. This is the problem for people who work with computer creativity. Computers are very good at reshuffling objects within the contexts provided by the programmer, but they cannot discover new contexts. Human beings can discover new contexts because of our nonlocal consciousness that enables us to jump out of the system. Moreover, we have access to the vast archetypal content of the quantum states of the mind (the pure mental states) that extend far beyond the local experiences within the lifetime of one person. Creativity is fundamentally a nonlocal mode of cognition.

The simultaneous discovery of the same scientific idea by people not connected locally in different times and places provides impressive evidence of nonlocality in creative acts.[3] This phenomenon is not limited to the realm of science. Similarities in the creative work of artists, poets, and musicians living in different times and places are so striking that they suggest nonlocal correlation as well. In this way, at least circumstantial evidence shows that creativity involves nonlocal cognition—a third way of knowing in addition to perception and conception.

THE CREATIVE ENCOUNTER

It is generally recognized that there are at least three distinct stages of the creative process.[4] The first is the preparation stage of information gathering. The second is the major stage of the creative process—germination and communication of the creative idea. The third and final stage is that of manifestation, in which form is given to the creative idea. I doubt, however, that creativity is the product of progressing through these three distinct stages in an orderly fashion.

Instead, I propose that the creative act is the fruit of the encounter of the self's classical and quantum modalities. There are stages, but they are all tangled-hierarchical encounters of these two modalities; the hierarchy is a tangled one because the quantum modality remains preconscious in us. The unitive consciousness is the inviolate level from which all creative action flows. Creativity is a tangled hierarchy because there is a manifest discontinuity even from the point of view of the classical modality.

The classical modality of the self, like the classical computer, deals with information, but the self's quantum modality deals with communication. Thus the first stage of the play of creativity is the tangled play of information (development of expertise) and communication (development of openness). It is tangled because you cannot tell when information ends and communication begins; there is a discontinuity. Here the ego acts as the research assistant of the quantum modality—and it takes a strong ego to handle the destructuring of the old that makes room for the new.

In the second stage, that of creative illumination, the encounter is between the perspiration of the classical modality and the inspiration of the quantum modality. In order to gain insight into this encounter, let us speculate about the details of the quantum mechanism—the details of the quantum jump in a creative insight. When the brain's quantum state develops as a pool of potentialities in response to a situation of creative confrontation, the pool includes not only conditioned states but also new, never-before-manifested states of possibility. Of course, the conditioned states of our own personal, learned memories are heavily weighted in the probability pool, and the statistical weights of the new, as-yet unconditioned states are small. Thus the problem of the second stage of creativity is this: How do we overcome the overwhelming odds that favor the artfulness of old memory over the genuine art of the new in this game of chance?

The answer to this is not all that obscure. There are five nonexclusive possibilities. First, we can minimize the mind's conditioning by consciously keeping an open mind to reduce the probability of (unconscious) conditioned responses. (This is also recommended for the first stage of creativity.)

Second, we can increase the odds of manifesting a low-probability creative idea by being persistent. This is important because persistence increases the number of collapses of the mind's quantum state relative to the same question—thus increasing the chance to realize a new response.

Third, since the probability of the appearance of a new component in the mind's coherent superposition is better with an unlearned stimulus (one to which we have not been exposed before), creativity is enhanced if we confront ourselves with unlearned stimuli. Thus reading about a new idea can trigger a shift of contexts in our own thinking about an unrelated matter. Unlearned stimuli that seem ambiguous—as in a surrealistic painting—are especially useful for opening our minds to new contexts.

Fourth, since conscious observation collapses the coherent superposition, there is a certain advantage to unconscious processing. Then uncollapsed coherent superpositions can act upon other uncollapsed coherent superpositions, thus creating many more possibilities to choose from in the eventual collapse.

And fifth, since nonlocality is an essential component of the quantum modality, we can enhance the probability of a creative act by working and talking with other people—as in brainstorming. The communication extends beyond the local interactions and the locally learned bases of the people involved, and the probability is high that the whole will be greater than the sum of the parts.

Thus, although the quantum modality performs the essential role of enabling us to make the jump out of the system that is necessary for the discovery of a truly new context (the inspiration), the classical modality performs an equally essential function: It ensures the persistence of the will (the perspiration). The importance of this persistence is noted by G. Spencer Brown in words that evoke the inexorable quality of what it means to have a burning question: "To arrive at the simplest truth, as Newton knew and practised, requires years of contemplation. Not activity. Not reasoning. Not calculating. Not busy behavior of any kind. Not reading. Not talking. Simply bearing in mind what it is one needs to know."[5]

The creative individual's ego has to be strong-willed to be persistent and has to be able to handle the anxiety associated with unknowing—with the quantum jump into the new. The contribution of the classical ego is justly recognized in the saying, Genius is 2 percent inspiration and 98 percent perspiration.

The third and final stage of the creative process, the manifestation of the creative idea, is the encounter of idea and form. The classical modality has the primary responsibility to provide form to the creative idea generated in stage two. It must sort out and organize the elements of the idea and verify that the idea works, but there is much going back and forth between idea and form. This interactive process happens in a tangled hierarchy.

Thus creativity is the tangled-hierarchical encounter of the classical and quantum modalities of the self: information and communication, perspiration and inspiration, form and idea. The ego has to act—but under the guidance of an aspect of the self that it knows not. In particular, it must resist reducing the creative process to a simple hierarchy of learned programs. Such reduction in the cause of efficiency is a natural but unfortunate tendency of the ego. The following lines from Rabindranath Tagore summarize all these aspects of the creative encounter:

> Melody seeks to fetter herself in rhythm,
> While the rhythm flows back to melody.
> Idea seeks its body in form,
> Form its freedom in the idea.
> The infinite seeks the touch of the finite,
> The finite its release in the infinite.
> What drama is this between creation and destruction—
> This ceaseless to and fro between idea and form?
> Bondage is striving after freedom,
> And freedom seeking rest in bondage.[6]

The Creative Ah-ha Experience

It is said that Archimedes, when he discovered the principle of buoyancy while in his bath, forgot his nakedness and dashed into the street rejoicing: "Eureka, eureka" (I found it, I found it). This is a famous example of the ah-ha experience. How is this experience to be explained?

The model of creativity as an encounter of the classical and quantum selves gives a succinct explanation of the ah-ha experience. Recall the time lag between the primary and secondary experiences. Our preoccupation with the secondary processes, indicated by the time lag, makes it difficult to be aware of our quantum self and to experience the quantum level of our operation. A creative experience is one of the few times when we directly experience the quantum modality with little or no time lag, and it is this encounter that produces the elation, the ah-ha.

The ah-ha experience typically occurs in stage two of the creative encounter; it is not the end, the product of the creative act. Stage three is a very important part of the process and consists of giving a manifest form to the creative idea that is germinated in the ah-ha experience.

So it seems Archimedes had a good dose of primary process experience that caused his ecstasy. I have already mentioned Abraham Maslow's work on peak experiences. What Maslow calls peak experiences can also be recognized as the creative ah-ha experience except that Maslow's subjects were not discovering a law of physics. Instead, they are examples of inner creativity—the creative act of self-realization.[7]

Outer and Inner Creativity

Understanding creativity as a common expression of the quantum self can encourage anybody to engage in it. In this context, we should distinguish between outer and inner creativity. Outer creativity involves discoveries external to oneself; the product of outer creativity is meant for the society at large. In contrast, inner creativity is inner-directed. Here the product is personal transformation of one's own context of living—a newer and newer *us*.

In outer creativity, the product we create competes with the existing structures of the society. Thus we need raw talent or giftedness and knowledge (including early conditioning) of existing structures in addition to a creative engagement with the problem that is to be solved. This combination may occur in relatively few people, although this scarcity does not have to be the case.

Inner creativity needs neither talent nor expertise. All it requires is a deep curiosity of an immediate, personal kind (What is the meaning of my own life?). All it needs is to recognize that with ego-development there is a tendency to neglect our creative power—especially in the matter of further self-development—and to say, in effect, I am who I am, I'll never change. All that inner creativity needs is to realize that the ego-level life, however successful, contains unease and lacks joy.

INNER CREATIVITY

The universe is creative; you and I in our creativity are the living proof of it. In determinism the world machine allows us to evolve only in its image, as mind machines. But there really is no world machine. In our desire for harmony and for prediction and control of our environment, we created the idea of the world machine and projected that deterministic image onto nature. A statically harmonious, lawful universe would be, however, a dead universe; the universe is not dead because we are not dead. We do, however, have the tendency toward a deathlike stasis: That tendency is the ego.

The Persian mystic Zarathustra is said to have laughed when he was born. Like many myths, this one has significance; it signifies that consciousness, as soon as it becomes manifest, is in a dilemma—laughable in its inability to escape conditioning. Only a baby can laugh at conditioning. By the time that baby reaches adulthood, it too will be conditioned—like everybody else—by society and culture, by civilization. Seeing a Woody Allen movie, we may very well conclude that neurosis is the price we pay for civilization, for societal conditioning; and Woody Allen's message is "dead" right. Chances are great that the grown-up child will be neurotically unable to laugh at her conditioned existence.

Even so, every now and then our creative nature breaks through our conditioning. Some of us have creative insights. Others radiate life on the dance floor. Still others find creative ecstasy in totally unexpected contexts. These are reminders. When creativity bursts through the ego, we get an opportunity to remember that there is something beyond the conditioned self. We may then wonder how to go about discovering what is beyond. How do we find a direct connection to the source of life-affirming meaning?

We are often quite fascinated by ourselves and our manipulations. Frequently this fascination intensifies in our teens. We become fascinated with our creative abilities, and we employ them to manipulate the world. This self-focused fascination continues a long time for many of us; for some it never stops. This fascination, moreover, is often productive and it has given us many wonders of our civilization.

But nothing is permanent in this world. Though I may have been high on creative juices yesterday, today a bite from the three-headed demon of universal afflictions may have filled me with ennui. The three heads of the demon are boredom, doubt (conflict), and pain.

What do we do when such suffering inundates us in the course of daily life? If we are still fascinated by ourselves, we cultivate escapes. In a sometimes obsessive flight from boredom, we pursue novelty—a new mate or a new video game—as a shield against that particular demon. To avoid the pain of discomfort, we seek pleasure: food, sex, drugs, and all that. And we secure ourselves in tight systems of belief as insurance to forestall doubt. Alas, all of these efforts are only more conditioning.

Trying to solve problems of inner emptiness and doubt with external fullness or internal rigidity is a materialist, classical approach. If we can change the world (and other people as part of that world), then we do not have to change ourselves. And yet, because reality is not static, we do change: We become cynical, or we slip into a mind-numbing hopelessness. We fluctuate between highs and lows, valleys and mountains, and life becomes a roller-coaster ride, a cheap melodrama, a soap opera.

Even our wonderful civilization, of which we are justifiably proud, threatens us in a big way. The creativity of our fellows that has provided our affliction-dodging toys of amusement has also delivered destructive toys that promise and deliver unquestionable suffering. This makes some of us wonder whether it is possible to be wisely creative. Can we use creativity to gain wisdom? Can we express creativity in ways that are constructive?

There is a story about Gautama Buddha: Once there was a very violent man in Bihar, India, where Buddha lived. This man, named Angulimala, had vowed to kill one thousand people. As a memento and a count of his victims, he severed an index finger from each victim and made a garland of fingers to wear around his neck (hence his name, Angulimala, which translates as "garland of fingers"). Pretty gory? Well, after his 999th kill, he fell prey to a slump (well known in sports circles—the problem of getting the record-breaking home run or winning the last leg of a tennis grand slam). Nobody approached near enough for him to claim his thousandth victim. Then came Buddha. Ignoring all warnings and pleadings, Buddha approached Angulimala. Even Angulimala was surprised that Buddha had come to him voluntarily. What kind of a man was this?

"Well, I'll grant you one wish for your bravery," Angulimala offered magnanimously.

Buddha requested that he chop off a branch from a nearby tree. *Whack*, it was done.

"Why did you waste your wish?"

"Will you grant me a second request, a dying man's request?" Buddha asked humbly.

"All right. What is it?"

"Would you restore that fallen branch to the tree?" asked Buddha with perfect equanimity.

"I can't do that!" exclaimed Angulimala, startled.

"How can you destroy something without knowing how to create? how to restore? how to rejoin?" asked the Buddha. It is said that this encounter so moved Angulimala that he became enlightened.

But the question Buddha raised two-and-a-half-thousand years ago remains relevant today. Suppose we ask our scientists, who use their creativity to invent weapons of destruction, the same question. How do you suppose they will answer?

Creativity unguided is a two-edged sword. It can be used to enhance the ego at the expense of civilization. One must apply creativity with wisdom, which leads to a transformation of being, so that we can love unconditionally or act altruistically. But how does one acquire wisdom?

No concrete specifications can describe what brings about wisdom or what exactly makes one wise. A Zen story makes the point this way: A master is asked by a monk to explain the reality beyond reality. The master picks up a rotten apple and gives it to the monk, and the monk is enlightened. The point is this: A heavenly apple of wisdom is perfection. The earthly apples of knowledge with which we comprehend the idea of transcendence are rotten apples, just confusing allegories and metaphors. However, that's all we have; they will have to do to get us started.

If you are able to handle the uncertainty of being beyond ego, you are ready for inner creativity. The methods of inner creativity include techniques such as meditation, which might be defined as the practiced attempt to achieve self-identity beyond the ego. Some techniques of inner creativity, such as Zen koans, use paradox explicitly. In other techniques, the paradoxes are more subtle.

One paradox is this: We use the ego to go beyond ego. How is this possible? For ages many mystics have marveled about this paradox of inner creativity, but it dissolves when viewed from the perspective of the new psychology of the self (chapters 12 and 13). Our self is not the ego. The ego is only an operational, temporary identity of the self. In attempting to weight our being more heavily toward the quantum modality, we recognize that we cannot force quantum jumps via any conditioned maneuver. So we systematically attack conditioning. We cannot gain more access to the quantum modality while constantly feeding the ego's demon-agent of affliction. So we give up some of our pursuit of pleasure, our attachment to excitement, our frantic attempts to avoid boredom, doubt, and pain. We give up limiting, runaway belief systems such as materialism. What happens? Are you ready to find out?

To put it in a different way, changes continually occur in our psyche as we accumulate experiences, but ordinarily these are low-level changes. They do not transform us. What we do in inner creativity is to direct the force of creativity specifically to the self-identity. Usually, creativity is directed to changing the external world, but when we creatively transform our own identity, it is called inner creativity.

In outer creativity, quantum jumps enable us to view an external problem in a new context. In inner creativity, the quantum jump allows us to break from established patterns of behavior, which together make up what is known as character, that have evolved through acts of growing up to adulthood. For some it involves a discontinuous ah-ha experience or quantum leap, such as a Zen satori. For others there is what seems to be a gradual turning about. It always involves patiently being aware of what is the immediate case, of what barriers are arising from our past conditioning that prevent us from living a new context that we intuit.

Remember Plato's cave? Plato characterized the plight of human beings in their experience of the universe in the following way: We are in a cave strapped in our respective seats, our heads fixed so that they always face the wall. The universe is a shadow show projected on the wall, and we are shadow watchers. We watch illusions that we permit to condition us. The real reality is behind us, in the light that creates the shadows on the wall. But how can we see the light, strapped as we are so that we cannot turn our heads? What was Plato saying with his analogy? And what about us, the people in the cave? We also cast a shadow on the wall, a shadow with which we identify. How do we loosen this ego-identity?

A modern-day Plato, Krishnamurti, suggests an answer.[8] We need to make a complete about-face, to transform, and this requires complete awareness of what is the case, of what we are, of what is our conditioning.

For example, suppose that you have a problem with jealousy. Every time your significant other talks to a person of the opposite sex, you are engulfed by intense pangs of self-doubt and anger. You try to change your feelings and behavior, but you cannot change by thinking or reasoning. This is where inner creativity comes in. The techniques of inner creativity are designed to create a slight gap between you and your ego-identification. In that gap, you have the ability to exercise your free will, the perfect right of your quantum modality.

So what does one do to achieve transformation? For outer creativity, we develop a talent or some expertise, or both—and yet creativity is not any of these things. Similarly, for inner creativity, one develops and practices awareness of one's conditioning—what is the case within. In outer creativity, if we have sufficient talent and have developed a certain expertise, then, if we are open and have a burning question, a creative quantum jump can happen. Similarly, in inner creativity, when we are aware of our inner-growth potential yet have no pretentions about ourselves, when we are vulnerable, then we can change. So, in either case, the doing is just the trigger. Both inner and outer creativity involve discontinuity and acausality.

How do we know that we have transformed? We know when the context of our living shifts from our personal ego level to the buddhi level, from the domination of classical self to a more comprehensive functioning in both classical and quantum modalities. What does that mean? In the simplest terms, it means a general condition of living with a natural sense of love and service to others—a natural surrendering of our separateness to the quantum self. Rabbi Hillel said,

If I am not for myself, who am I?
If I am only for myself, what am I?

When both questions initiate our actions with equal urgency, then there is transformation. However, transformation is an ongoing process, always defining an ever-more-compassionate context for our being.

Stages of Adult Development

Of all cultures, perhaps the East Indians have done the most extensive research on inner creativity. One of their findings, which now is being confirmed by science, is the developmental nature of inner creativity. The Hindus delineated four developmental periods for students of inner creativity:

1. *Brahmacharya* (which literally means "celibacy")—a period of learning and ego-development, including some initiation into spirituality, covering childhood and young adulthood.
2. *Garhastha* (which literally means "living as a householder")—a period of living in the world with ego-identity and enjoying the bittersweet fruits of the world.
3. *Banaprastha* (which literally means "dwelling in the forest")—a period of looking inward and cultivating the awakening of buddhi.
4. *Sanyas* (which literally means "renunciation")—a period of development in buddhi leading to a renunciation and transcendence of all dualities, of all the various drives, and thus to liberation.

The current paradigm of psychology universally acknowledges only the first two of these developmental levels. However, a few researchers—notably Erik Erikson, Carl Rogers, and Abraham Maslow—have suggested a broader context for development of the human being.[9]

Also noteworthy is the idea of midlife transition popularized in the seventies. Obviously, that formulation touched many people, as implied by the following joke: A priest, a protestant minister, and a rabbi were discussing the point at which life begins. The priest gave his standard answer, "Life begins at conception." The minister equivocated: "Maybe life begins after twenty days or so?" Finally the rabbi said: "Life begins when your kids are gone, and the dog dies."

In the following chapter, I will examine, in accordance with the idealist literature and with the insights explored in this book, the idea of the awakening of buddhi. The further stage of maturation in buddhi that leads to the freedom called *moksha* in Hinduism and *nirvana* in Buddhism is highly esoteric and is beyond the scope of this book.

Chapter 17

THE AWAKENING OF BUDDHI

IN ONE OF THE UPANISHADS appear the following evocative lines:

> Two birds, united always, and known by the same name, closely cling to the same tree. One of them eats the sweet fruit; the other looks on without eating.[1]

This is a beautiful metaphor for the two ends of the spectrum of the self; at one end we have the classical ego; at the other, the quantum atman. In our ego we eat the sweet (and bitter) fruit of worldly pleasure and seem oblivious of our quantum modality, which gives meaning to our existence. We externalize ourselves in local pursuits and become lost in the usual worldly dichotomies—pleasure and pain, success and failure, good and evil. We little heed the possibilities available to us in our internal nonlocal connection, except perhaps for an occasional foray into creativity and conjugal love. The older we get, the more stuck we become in our ways. How do we change this modus operandi and develop an individual program of adult development?

Fortunately, much empirical data have been gathered over millennia and summarized in spiritual literature. Before we get into a discussion of these strategies, it is necessary to understand the metaphor of the two birds.

Many people think of the spiritual journey as analogous to mountain climbing and the different spiritual paths as paths up the sides of the mountain. In this way of thinking about the metaphor, there is a tendency to think hierarchically and to assume that since we seem to be seeking a goal (the mountaintop), then the closer we are to the top, the better we are. Once again we are caught in an ego-level superiority-inferiority dichotomy.

The opposite is to pronounce, like the mystic Krishnamurti, Truth is a pathless land. But if there is no path, very little guidance can be given. This is a tremendous waste of the wisdom gained from the available empirical data.

Yudhisthira, one of the heroes in the ancient Indian epic, *Mahabharata*, was asked under the threat of death the following question: What is religion?

Yudhisthira's reply, which saved his life, is worth remembering: "The maps of religion are hidden in the cave," said he. "Studying the ways of great men and women reveal the path."

So we shall consider paths to be examples of the kind of methods that have been used in the past and still are used to shift our identity from the ego level through buddhi toward atman.

According to the Bhagavad Gita, there are three major paths, each of which is called a yoga. *Yoga* is a Sanskrit word meaning "union" (etymologically, the English word *yoke* has the same origin). Here is a further meaning of our metaphorical two birds: The birds are already united. The task of yoga is to recognize the union. The recognition begins the shift of the identity.

The three yogas emphasized in the Gita are as follows:

1. *Jnana yoga*, the path of illuminating the intellect with intelligence (buddhi). (*Jnana* is the Sanskrit word for knowledge.)
2. *Karma yoga*, the path of action in the world. (*Karma* is the Sanskrit word for action.)
3. *Bhakti yoga*, the path of love. (*Bhakti* is the Sanskrit word for devotion, but the spirit of the word is quite close to love.)

These three yogas are by no means unique to the Gita or to the Hindu tradition. Jnana yoga is popular in Zen Buddhism. Catholicism tends to favor karma yoga (the ability to effect transformation through the actions known as sacraments), and Protestantism leans heavily toward the path of love. (The love of faith is reciprocated by the love known as grace, but grace cannot be merited by action.)

Jnana yoga intends to awaken the intelligence of buddhi by using the intellect, but the trick is to trigger a change of the usual contexts in which the intellect works. Intellect is an artful caricature of creativity; it involves a reasoned reshuffle of known contexts; it is creativity mixed with the other ego-level drives of conditioning and libido. How can we arouse the intellect into comprehending a new self-identity? If you asked this question of a Zen master, he might clap his hands and ask you to hear the sound of only one hand clapping. The clap is to startle the Upanishadic "bird" lost in illusion, to make it jump—a quantum jump to realize its union. A paradox is a very effective way to stir up a stuck intellect. A person thinking about a paradox enters a double-bind situation and must take a leap to escape the double-bind. The technique is commonly used in Zen Buddhism.

There is much misunderstanding about Zen koans. They often seem so pointless. Once at a party I met a fellow who had recently returned from Japan, where he had spent some time at a Zen monastery. He asked the koan, What is the sound of one hand clapping? Several people at the party became frustrated in their attempt to solve the puzzle. After all, how can one clap with only one hand? It takes two hands to clap, doesn't it? Finally, the fellow gave in and demonstrated his solution. He hit his hand on a table. That was the sound of one hand clapping. Everybody at the party was delighted.

It is easy to regard koans, as this man did, as merely puzzles to be solved intellectually, and they can be fun to investigate rationally because they lend themselves to all sorts of imaginative possibilities. But such purely intellectual solutions will not help us to lift the veil that the ego represents. The function of a koan is much more subtle. If we tried the table-clapping solution of the above koan with a Zen master, the master might say, I'll hit you thirty times (or he might do it), or give you a score of 20 percent, or make some other equally unprofound response. He would know that you had not gotten the koan.

In our ego we are impatient to know the answer to puzzles and paradoxes rather than to understand their meaning. We intellectualize rather than intuit. Intellectualization alone simply reinforces the ego's inertia. It has its place, but at the proper moment the intellect must surrender to unknowing so that new knowledge can enter.

This point is made very powerfully in a Zen story. A professor came to visit a Zen master with the idea of learning something about Zen. The master asked if the professor would like some tea.

As the master was preparing the tea, the professor started expounding his knowledge of Zen. The tea was made, and the master started pouring the tea in the professor's cup; the cup became full, but the master went on pouring.

The professor cried, "But the cup is full!"

"So is your mind with ideas about Zen!" admonished the Zen master.

The anthropologist Gregory Bateson noted the similarity of the koan technique to the double-bind.[2] The double-bind neutralizes the ego by paralyzing it. The ego-self cannot handle the no-win oscillation from one option to another in such a transaction as this: If you say this dog is Buddha, I will hit you. If you say this dog is not Buddha, I will hit you, and if you don't say anything, I will hit you.

The imperative conditions that create a double-bind are that (a) two people are involved and (b) there is a bond between these two people that cannot be broken. That is, the situation is such that the person in the double-bind has temporarily surrendered his ego autonomy. Of course, once the jump to a new context of living takes place—an event called *satori*—the job of the master is accomplished, and he lovingly releases the double-bind.

The Zen master targets the thinking mind for the double-bind catapult to transcendence of the ego-identity. Teachers in the Christian and Sufi traditions, in contrast, zero in on the feeling mind with their injunction to love without expectation. The ego-"I" itself is as unable to love unconditionally as to solve a koan. In both cases it is the creative quandary that the teachers want to intensify in their students.

Can you imagine loving someone from choice—not because there is the possibility of ego-gratification, not because you are in love, not because you have reasons to love? This is love from the level of buddhi. We cannot will it. We can only surrender to it in a creative opening.

There is a Chinese fable about the similarity and difference between heaven and hell. Both heaven and hell are banquets with large, round tables that are laden with delicious food. In both places, the chopsticks are about five feet long. Now for the difference. In hell people try in vain to use the chopsticks to feed themselves. In heaven everybody simply feeds the person sitting across the table. If I feed another, will I be fed? Surrendering this ego-level uncertainty is the awakening of trust.

Just as unconditional love demands trust from the lover, so it invites trust from the receiver. The great Chinese Taoist teacher Chuang Tzu used to tell his students this parable: Suppose that a man is traveling in a boat and suddenly sees another boat coming at him. Reacting with irritation and anger, he shouts loudly and gesticulates madly to the helmsman of the boat to change his course. But then the boat comes closer, and he sees that there is no one in it. His anger dissipates, and he himself now steers clear of the empty boat.

What happens, asks Chuang Tzu, if we approach others from an emptiness of the heart, without preconceived ideas? In that unbiased emptiness, the probability pool of choice is extended to the creative dimension. The quantum wave of our mind expands and is ready to embrace new responses: I am not driven to love by desire, by a need for security, by image, but I am free to love for no reason at all. It is this unconditional love that conquers our reactivity.

Of the three yogas emphasized in the Bhagavad Gita, karma yoga is at once the most elementary and the most difficult. It is also the most urgent for our age, for appropriate action is the final goal of karma yoga. On the way to the exalted being from which appropriate action flows easily without effort, one must acquire much spiritual development. The Gita suggests a threefold, step-by-step approach.

The first step is to practice action without coveting a particular fruit of the action. "Give the fruit of the action to God," says the Gita. This is what normally is called karma yoga.

In the second stage, one acts in the service of God. If you ask Mother Teresa where she finds the stamina to serve the destitute in Calcutta and throughout the world day-in and day-out, she will say, I serve Christ by serving the poor. She daily meets Christ in her work, and this is enough for her. This is karma yoga in which love has awakened.

At the final stage, one lives as the agency of appropriate action—not as the subject acting on an object. This is karma yoga at the point of liberation.

Although spiritual development occurs in stages, no method is restricted to only one stage. All three yogas—action, love, and wisdom—are employed simultaneously at all stages of self-development. In Buddhism, one explicitly recognizes this spiral nature of the different yogas. If you look at the Buddha's eightfold way, you will find within it all three paths. We use them together, each path enhancing the other. The more we act without the fruit of the action, or the more we meditate, the more able we are to love. The more we love, the more mature becomes our wisdom. The wiser we are, the more natural is selfless action.

Notice that all these paths depend on our being aware of what is going on inside and outside of us. This awareness is so crucial to all paths that when Krishnamurti says there is no path and advises only awareness, he is right. All one needs is the practice of awareness, which is meditation.

Jnana: Waking Up to Reality

When we connected mysticism with monistic idealism (chapter 4), we introduced the concept of consciousness as the ground of being, Brahman. As we developed a cosmology about how the one becomes many, it became clear that the Brahman consciousness arises as the subject (atman) co-dependently with objects. Arising co-dependently are the knower (the subject of experience), the field of knowledge (awareness), and the known (the object of the experience). There is, however, no self-nature, no independent existence, in either the subject or the object: Only consciousness is reality.

The problem is how to comprehend this reality. Language is inadequate here. Try, for example, There is only one consciousness. Good, up to a point, but by saying "one," we have already made a distinction, subtly implying duality. Thus the beautiful saying of Shankara: one without a second. Better, but not perfect. Another approach is conveyed by a joke: How many Zen masters does it take to screw in a light bulb? One and not one.

It is very difficult to express the non-relative reality in relative words. In his writings, which have been called the first truly post-modern philosophy, Jacques Derrida has introduced the concept of deconstruction—the undermining of all metaphysical statements about reality by undermining the very meaning of statements in general. Millennia ago, the Buddhist philosopher Nagarjuna suggested the same thing. The direct wisdom arrived at by intense practice of this deconstruction is the height of jnana yoga.

The quantum physics of self-reference now provides an additional way to ponder this imponderable: tangled hierarchy. There is nothing manifest before consciousness collapses the object/awareness in space-time. But without awareness, there is no collapse, no choice to collapse. What is before collapse? The tangled hierarchy—the infinite oscillation of yes-no answers—does not allow us to experience the original: the sound of one hand clapping. What is atman experience? To transform creatively the intellectual understanding of the idealist metaphysic into realized truth, go deep into the question—make sure in your gut, awaken your heart.

The mystic philosopher Franklin Merrell-Wolff said, "Substantiality is inversely proportional to ponderability."[3] This is the key hint in jnana yoga; the more imponderable it is, the more substantial it is. Follow the thought to subtler and subtler depths. Then. . . .

The aftermath is an awakening that leads to the buddhi level of self-identity. For most people, except for the occasional, rigorously trained scientist or philosopher, jnana yoga may seem too difficult. Fortunately, the other two methods (karma yoga and bhakti yoga) are more accessible to many people.

Meditation

According to many philosophers, there is only one method of inner creativity—meditation (which is learning to pay attention, to be detached, and to be a witness to the ongoing melodrama of thought patterns). To break away from the ego-level of existence, you may need to identify with some precision what is going on in your everyday life, to recognize, perhaps painfully, how your habit-attachments run you. Or to open to love, you may focus attention on your relationships in the world. Or you may want to contemplate reality. All of these techniques require basic practice in being attentive and in being detached. Meditation teaches us that.

Of the many forms of meditation, the most common is performed while sitting. If you keep your attention on your breath (eyes closed or open) or on the flame of a candle or on the sound of a mantra (usually performed with closed eyes) or on any object, you will be practicing concentration meditation. In this practice, whenever your attention wanders and thoughts arise, as invariably happens, you gently and persistently bring your attention back to the focus, maintaining a one-pointedness in order to transcend thought, to shift it from the foreground to the background of awareness.

In another form, called awareness meditation, thought itself—indeed, the entire field of awareness—becomes the object. The principle here is that if attention is allowed freely to observe the train of thought without engaging with any particular thought, it will remain in a resting state with respect to the moving thought-parade. This form of meditation can afford you a detached, objective view of your thought patterns that eventually will enable you to transcend thoughts.

The difference between concentration and awareness meditation can be understood by invoking the uncertainty principle for thought. When we think about our thinking, either the individual thought (the position) or the train of thought (the momentum) becomes fuzzy or uncertain. As the uncertainty about the individual thought becomes progressively smaller and smaller, the uncertainty in the train of thought tends to become infinite. With association gone, we become centered with the here-now.

In awareness meditation, it is the uncertainty in the association that is made progressively smaller and smaller, thus causing us to lose the feature or content of thought. Since attachment results from the content of thought, if the content is gone, so is the attachment. We become detached observers of, or witnesses to, our thought patterns.

MEDITATION RESEARCH

Do meditation techniques, absurdly simple in concept though quite challenging in practice, actually enable people to achieve altered states of consciousness? Brain physiologists, on the premise that there may be a unique physiological state corresponding to the meditative state of consciousness, have attempted to answer this question by measuring various physiological indicators (heart rate, galvanic skin resistance, brain-wave patterns, and so forth) as a subject meditates. Although this premise has never been validated, experienced meditators show such significantly distinct physiological characteristics that meditation has been recognized by many researchers as the fourth major state of consciousness (the other three are wakefulness, deep sleep, and rapid-eye-movement or dream-associated sleep). The main evidence for meditation as a distinct conscious state comes from EEG brain-wave studies.[4]

The brain-wave pattern of waking consciousness is dominated by low-amplitude, high-frequency beta waves (greater than 13.5 Hz). In meditation, these waves are replaced by high-amplitude, low-frequency alpha waves (7.5–13.5 Hz). This alpha dominance, which signifies a relaxed, passive receptivity, is one of the important characteristics of the meditative consciousness, although alpha dominance alone cannot be regarded as indicative of a meditative state. You can generate a predominantly alpha brain-wave pattern simply by closing your eyes.

Other striking characteristics of the meditative brain-wave pattern have been found. When people in an ordinary alpha state are subjected to a sudden stimulus, they respond by a sharp return to the beta mode. This phenomenon is called alpha blocking. In contrast, veterans of concentration meditation exhibit the uniqueness of their alpha pattern by showing no alpha blocking when a sudden stimulus occurs while they are in their meditative alpha state.[5] People who practice awareness meditation do display alpha blocking, and the uniqueness of their type of meditative alpha state reveals itself in a different way. A person in ordinary waking awareness, when exposed to a repeated stimulus (like the ticking of a clock) adjusts to the stimulus in a very short time to the extent that his brain-wave pattern no longer changes. This is called the habituation response. (It takes only four ticks of a clock to habituate a normal subject to the ticking.) Veterans of awareness meditation uniquely show no sign of habituation in either their meditative state or their waking state.[6]

Research has shown the importance of passivity of visual attention (the so-called soft eye) for generating the meditative alpha state. Such passivity may be achieved simply by tilting the eyes upward or downward, as is common in some Tibetan practices. High alpha is also achieved by passive attention to space.[7] It is now generally recognized that the alpha state is good because it generally signifies a relaxation from the tensions of the body and of the mind, thus freeing us to go deep into investigating the self.

Another aspect of the meditative state is the appearance of theta waves (3.5–7.5 Hz) in the EEG pattern. The theta waves may be highly significant because they are known to be associated also with the creative experience.[8]

The presence of theta waves in the brain-wave pattern of meditators brings to mind that young children up to five years old show theta dominance, which evolves into the alpha dominance of adolescents' normal waking brain-wave pattern and eventually is replaced by the beta pattern of the adult. Since children's developing consciousness has a dominance of the quantum modality (that is, less of the secondary-awareness processes), we can speculate that theta waves characterize in some way the primary processes of the quantum modality in the brain-mind. If this speculation is valid, both seated meditation and creative experiences, by their theta signature, may be showing a shift of consciousness to the primary process of the quantum mode.

Current research on attention is giving us a hint about how the mantra or concentration meditation works. In experiments by University of Oregon psychologist Michael Posner and his collaborators, subjects are given a single prime letter, such as *B*, followed after a variable interval by a pair of letters.[9] In some trials, the subjects are asked to pay attention to the prime. In others, they are not. The subjects respond yes or no, depending on whether the letter-pairs consist of identical letters, such as *BB*, and the reaction time taken for a response is measured.

The most interesting result, from my point of view, occurs when the subjects are told to pay attention to the prime letter in trials in which the subsequent letter-pair does not match the prime: There is a distinct cost of reaction time in these trials. Attention to the prime affects the processing of an unexpected item. (Conversely, if conscious attention is not given in these experiments to the prime, the reaction time remains unaffected.)

Thus the result of attention is to interfere with our ability to perceive objects that are different from the object of our attention. The quantum state of the brain develops in time as a probability pool that includes new stimuli, but focused attention on an existing stimulus biases the probability of response in favor of that stimulus, while the probability of collapsing a new perception becomes low. Attention to a mantra, therefore, diverts our attention from idle thoughts. Our consciousness literally cannot focus on two things at the same time. The external world that exists in us as an internal map starts giving way as we become better in attending to the mantra. Eventually, a state is reached where the thinking mind itself seems to habituate away: That is, although secondary-awareness events are present, they are few and far apart. This occurs when primary processes can reveal themselves in their suchness.

In awareness meditation, the strategy is likewise commensurate with our brain structure. After all, the thoughts and feelings of our secondary-awareness events are inevitable. We are unable to battle them for any prolonged period of time simply because of our brain structure. In awareness meditation, one allows this, but a distinction is made between the contents of consciousness and the subject:

consciousness itself. In the mystical literature the metaphor of muddy water is used to convey the idea:

> The seed of mystery lies in muddy water.
> How can I perceive this mystery?
> Water becomes still through stillness.
> How can I become still?
> By flowing with the stream.

If we flow with the stream, the so-called muddy contents of consciousness—our thought patterns—are consigned to the stream bed, to the bottom of the awareness that we witness. Using this strategy, we can be witnesses for longer and longer periods of time because we are no longer interfering with the secondary awareness experiences through introspection. This enables us to experience the suchness, or no-selfness, of the witnessing consciousness.

Thus in both concentration and awareness meditation, the suchness of the experience is amazing, because this suchness gives us a glimpse of a primary consciousness that is beyond the murmurings of the secondary ego. There is consciousness beyond thought and besides thought, beyond ego. The experience of this inner primary consciousness can be increased with practice.

Freedom in Meditation: Karma Yoga

The path of action, karma yoga, begins with the practice of learning to act without attachment to the fruit of one's action. The ego wants fruit. This is why the reward-punishment system shows up so universally in all cultures. Renouncing the fruit of the action is heretical to the habit-bound ego and, because of the renunciation of sanctions involved, to figures of authority.

So, the path of karma yoga involves renouncing the rewards and punishments that condition our behavior. How do we break with our conditioning? The answer is through meditation, which is part and parcel of karma yoga.

When you first start meditating, it is likely that nothing much will happen. It is a challenge to keep up the twenty or so minutes of sitting during this initial period. It calls for real discipline. In my own case, it took several months before I noticed anything.

Maggie and I began our marriage with a commitment to open

communication. In nonglorious terms, that meant for us that we fought a lot in the early years. After a fight, I usually suffered from negative thoughts dominated by trade-offs and back-ups—I'll show her, and all that. After I had meditated for about three months, I was upset one day after a fight, and yet I noticed that the usual negative thoughts about my wife were missing. Something had dropped.

Another time soon after, I was in a heated argument with my teen-aged stepson, who is also very logical, like me—and you know how irritating logic is during a battle of temper. I was angry, but suddenly I noticed that the anger was on the surface. Inside, I was enjoying his prowess in fighting back. I had the choice of reacting angrily or of enjoying the situation, and I was exercising my choice of saying no to the habitual reactivity. At first I mainly exercised this choice internally, but eventually it became manifest in my outer action as well.

Such incidents as these are actually quite common and can encourage us during the first few crucial months of practice. Most importantly, they show that meditation does help us to see the patterns of the ego. A few of them can even drop.

Pat Carrington, in her book *Freedom in Meditation*, relates how one of her clients gave up smoking: "While traveling on a plane he was meditating and had the impression that he heard his own voice saying: 'Empty yourself of your desires!' This rather mysterious statement was followed by an experience of exultation and further words: 'I can . . . smoke a cigarette if I want to—but I don't have to.' "[10]

What we aim for in meditation is to reduce our near-100 percent probability of a fixed response to a conditioned stimulus. For instance, I have the desire to smoke. The ego has two responses: I must smoke because . . . and its polar opposite, I shouldn't smoke because. . . . Meditation breaks the monopoly of these responses and opens a gap. In that gap is born the creative response of the above anecdote: I choose to smoke or not to smoke. Only when such a thought enters *creatively* can the radical change from smoker to nonsmoker occur. Such an event is possible when one's practice is intense and persistent.

The important thing is not to isolate meditation from the rest of your life but to allow it to transform your actions. You will find that this is not as easy as it sounds. The ego is well defended against change. The psychologist Richard Alpert (Ram Dass) tells of an occasion when he and a few friends had just completed a group meditation. Everybody was supposedly feeling content when one of the meditators, intent on having his cake and eating it too, said, "Oh, that was great. Now we can go out for some beer and pizza." It is quite a challenge to give up such compartmentalizing patterns.

After all, the idea that beer and pizza are good times and meditation is work is only a belief. As long as we maintain such beliefs, attentive seated meditation (no matter how blissful) is of little benefit. We must supplement our meditative practice with continuous, rigorous examination of our constraining belief systems. The idea is to practice, in the spirit of Mahatma Gandhi, not clinging to any beliefs that we do not fully live. Beliefs held but not practiced are vain. They are dead reflections of a passing show.

Einstein was sitting for his portrait with artist Winifred Reiber. He commented that Hitler, in prewar Germany, was hurting himself in the eyes of the world by confiscating the possessions that the Einsteins had left behind when they immigrated to the States, but Einstein's wife had a different perspective. She reminisced longingly about the personal treasures she had enjoyed in Germany and regretted that here she had so little. She missed "the silver, the linens, the rugs, the books and her grandmother's old Meissen ware." She was attached to those things. "But they were not attached to you," Einstein quipped.[11]

This is the thing. Our thoughts, our beliefs are not attached to us. They fall away if we do not hold on to them. Recently, the movie *Gandhi* inspirationally swept the world. I hope that the message of Gandhi has been received by a substantial number of us. Gandhi used to say, "My life is my message." He lived his beliefs. Any belief not lived is empty baggage. The purpose of meditation is to help us drop the baggage so that we can live freely.

Somebody once asked me during a seminar how I could preach giving up belief systems while at the same time helping to create a new idealist science that, in a sense, is also a belief system. This is a legitimate question to which I respond, in the Gandhian spirit: Do not make the new science a new belief system. Use it, or the philosophy of monistic idealism, or any of the teachings of the great traditions to shed existing belief systems that merely shackle your mind and heart. If you have the appropriate resources, join the new science's endeavor in support of enlightened living. Then the science will be your sadhana (practice), as it is for me. But if science is not your medium, and if you are committed to radical change, find your own path. Follow the path of your heart. Don't pick up someone else's baggage, or you will find the spiritual journey onerous under its weight.

THE AH-HA EXPERIENCE OF INNER CREATIVITY

The poet Rabindranath Tagore wrote:

> *Jewel-like the immortal*
> *does not boast of its length of years*
> *but of the scintillating point of its*
> *moment.*[12]

The secret of immortality is to live in the present moment, in the here-now; here-now is timeless. Like poets who glimpse immortality, teachers of inner creativity constantly talk about the importance of experiencing the here-now. But what exactly is meant by *here-now*? Most of us cannot grasp even intellectually, except as a sanitized abstraction, the meaning of the term—let alone experience this state of present-centeredness.

We cannot will habitual life in the here-now, but we can cultivate the conditions that allow such a life to come into being. With meditative practice—sitting and repeating a mantra or practicing choiceless awareness meditation—we can fall into it. The mantra can take us to the here-now by depriving our senses of any other stimulus but the mantra, freeing us to establish a new relationship with reality.

Here-now being is called samadhi when there is complete absorption with the object of meditation. The subject recedes to mere implicitness. In higher samadhis, the essence of the object is penetrated, and eventually the object is seen in its suchness, in its identity with the whole of consciousness. This is also called the experience of no-self because there is no particular self anywhere. Zen Buddhists call it satori marked by a vivid awareness of the suchness (tathata) of an object. Some people call it gnosis, or enlightenment. The state of samadhi or satori is accompanied by a feeling of intense joy.

A somewhat different experience of timelessness is when we achieve, through meditation, the state of perfect witnessing. Objects rise and fall in our awareness, but the witness is completely unattached, without judgments.[13] The experience produces the same effect—joy—as an aftermath. (Of course, the creative force of the experience is manifest only when we are eventually able to carry the perspective of witnessing into everyday life.)

The joy of meditative experiences is the original joy of consciousness in its pure form. In Indian philosophy, *Brahman*, the ground of being, is said to be manifest as *sat-chit-ananda*, where *sat* means existence, *chit* means consciousness, and *ananda* means joy. Everything that is manifest in space-time is sat. Things exist. In contrast, self-consciousness is very special. It needs a brain-mind in order to be manifest. Joy is even more special. It takes the self after an ego-development to recognize that it is experiencing something much greater than the individual self. This recognition produces joy—the joy of the glimpse of who we really are.

Some traditions also call this ah-ha experience of inner creativity enlightenment. There is some appropriateness in this name. In our ego we tend to identify ourselves with our brain-mind. In samadhi, we recognize that our identity is in the light of consciousness that infuses us and all of existence. The ego has no substance.

Unfortunately, the term *enlightenment* also creates a lot of confusion. Many people conceive of the enlightenment experience as an accomplishment: I am enlightened now. Although the experience has opened the door for the shift of self-identity, the ego-level tendency continues, and the accomplishment orientation may thwart complete transformation.

But the experience itself is only the threshold to this transformative potential. A creative act is incomplete without a product, and inner creativity is no exception. After the ah-ha experience of samadhi or satori or perfect witnessing, a disciplined practice is still needed to translate the awakening of buddhi into action in the world.

The Awakening of Love: Bhakti Yoga

In the Bhagavad Gita, Krishna makes a highly revealing comment to Arjuna. Arjuna, he says, I will tell you the secret of all secrets, the most straightforward way to awaken buddhi. It is to practice seeing Brahman (translate *Brahman* as God in this context) in everything and everybody and to serve Brahman as a devotee. There is no need to struggle with subtle forms of discursive wisdom. There is no need to practice action without the fruit of the action. There is not even the need for formal meditation. Just love God and serve God in everyone. (It's a little like getting the card in the game of Monopoly that says Go directly to Boardwalk.)

Of course, here, too, there is subtlety. What does it mean to love God? Many people misunderstand. They think it is to develop a relationship of ritual worship to some idol or idea of God.

The idealist literature points out five ways to love God, all involving a human form:[14]

1. Loving God through loving self
2. Loving God through service
3. Loving God through friendship
4. Loving God through the mother-child relationship
5. Loving God through an erotic relationship

The list is not exclusive. There are other very tangible methods. For example, Francis of Assisi practiced loving God through loving nature—a practice that today is forgotten in Christianity but lives on in the native American tradition. Imagine what resurrection of this practice would do for our environmental causes.

What one tries to do in the method of love is, first, to escape the dominance of locality in our relationship to nonlocal consciousness. Certainly in every human relationship, locality dominates. We communicate through sight, sound, smell, touch, and taste, the usual sense experiences. But these are not the only means of communication. If they were, it is doubtful that we could really communicate meaningfully to one another. So we practice devotion to the spirit of relationship, giving up legalistic score-keeping in our transactions with others.

Second, as mentioned before, the ego becomes a solipsistic universe for each of us, a locked prison cell in which only I and my extensions are real. Others have to defer to me, my culture, my race, and so forth, to be acceptable in my universe. Developing unselfish love relationships is one way—perhaps the most direct way—to break through the ego's solipsism.

The ego loves itself, so much so that it wants to be immortal. This seeking of immortality expresses itself in the West in the striving for fame and power. In the East it has led to the idea of reincarnation of the individual soul. Can this love of the ego turn to a love of the atman: the inner quantum self? One has to discover a different immortality. Through love, through patient forgiveness of self and others, one focuses on the permanent aspect of oneself as a way of transcending the transitory ego. This method is called *santa* in Sanskrit and translates as "passive." It has been common in many contemplative Christian communities.

The other four ways in the list above involve active participation in relationships with others. Altruistic service to others, called *seva* in Sanskrit, comes naturally to many people—a fact that confounds the proponents of the selfish-gene idea, who believe that altruism is possible only if there is a common genetic inheritance among the people involved. Seva is the practice of Mother Teresa, who serves people as the expression of her love for Christ, and what a glorious expression it is. Service involves sacrifice of selfish needs and desires, which is a direct affront to the ego's solipsism. When love breaks through, that marks the awakening of compassion—and compassion is an essential ingredient of Soto Zen practice.

In America, we have almost lost the institution of friendship among males because of the myth of the value of rugged individualism and because of the marketplace-based economic model of relationship. In that model, one evaluates relationships by means of a cost-benefit analysis. Fortunately, the tendency to apply such pragmatic criteria to friendships may be reversing somewhat, if the popularity of poet Robert Bly's recent work on male bonding is any indication. Another major difficulty with friendship in America is the demand for efficiency. Friendship is not always efficient. It often involves self-sacrifice, suspending efficiency and time constraints, breaking through the ego's cocoon. Women in America have traditionally been less bound by the market-economy model of relationships. Currently, however, pressures in this direction are growing as more women work in the marketplace and try to stretch their time and energy to meet the demands of both career and home. If they can resist this pressure, perhaps women will bring their capacity for loving friendships into the marketplace with them and so teach men how to humanize their economic interactions and how to be friends again.

The Man-Woman Relationship

Because of biological differences, intimacy is a unique challenge in the man-woman relationship and has great potency for breaking through the ego-boundaries.

An intimate relationship with another of the same sex is in a sense easier because of the common gender experiences we share with the other person. But a man and a woman, subject as they are to different biological and social conditioning, practically belong to two different cultures. In terms of the Jungian archetypes (anima, the female experience suppressed in man, and animus, the male experience suppressed in woman), a consequence of the requirements of form is suppression that causes a deep chasm in our ability to communicate with the opposite sex.

There is a mythical story in Plato's *Symposium.* Originally, humans existed as bisexual creatures with two sets of arms, legs, and sexual organs. But the power of these bisexual creatures was so great that the gods feared a usurpation of the prerogatives of heaven. Therefore, Zeus cut the creatures into two. Thereafter, the divided humans forever sought their missing halves. This story metaphorically captures the unconscious drive we have to make the unconscious archetypes of anima or animus conscious so that we can become whole. But the unconscious drive is not only instinctual, it is also Freud's eros of the personal unconscious. Eros is augmented by creativity from the collective unconscious.

Somewhere along the path to intimacy between two committed people, the anima in man and the animus in woman are aroused, and both may become empowered into the buddhi level as a result. Think about it. The reason I am solipsistic in my ego is that there really is no local way of putting myself in somebody else's shoes. (Read the article "What is it like to be a bat?" by Thomas Nagel.[15]) So, my tendency is to think that my private universe is universally representative. The anima and animus experiences are true nonlocal experiences, and suddenly otherness makes sense—the other becomes a human being like me. His or her individual experiences and perspectives become as valid as mine. When we discover this otherness, we have discovered unconditional love—love that can catapult us to the buddhi-level of being.

Once we have broken through the cocoon of our ego-solipsism with even one person, we have the potential to love others intimately. It is like extending your family. That's why the Sanskrit proverb says that "to the liberated, the whole world is family."

As the whole world becomes family, we begin to see the true nature of the immanent consciousness. We see the unity in diversity. We love people for their being. We do not need or want them to conform to our particular patterns and cultures. Instead, we respect them and marvel about the scope and extent of diversity. We begin to see what the Hindus call God's play, *lila.*

The flute of interior time is played whether we
hear it or not.
What we mean by "love" is its sound coming in.
When love hits the farthest edge of excess, it reaches a wisdom.
And the fragrance of that knowledge!
It penetrates our thick bodies,
it goes through walls—
Its network of notes has a structure as if a million
suns were arranged inside.
This tune has truth in it.[16]

Chapter 18

AN IDEALIST THEORY OF ETHICS

THE CHARACTERS Ivan and Alyosha in Dostoyevsky's unforgettable novel *The Brothers Karamazov* are obsessively torn by ethical considerations of right and wrong, but this was written in 1880. How often do modern men and women give such significance to ethics in their actions? The implicit adoption of a cognitive science-behaviorist view of ourselves—the idea that we are classical machines and therefore determined by our genetic and environmental conditioning—has played a major role in eroding the importance of ethics and values in our society. Our moral values are too often filtered through the hypocrisies of political pragmatism and a rationalizing that honors the letter over the spirit of the law. We greedily conform to consumer-exploitive images of the good life. In such a culture, traditional values are a broken rudder with little strength to steer a meaningful course among the choices, large and small, that can shipwreck us.[1]

Similarly, we lack any strong guide when we attempt to focus on the ethical dimensions of scientific and technological projects, such as genetic engineering and the arms race. Can we ever justify ethics scientifically? Can we find a scientific basis for ethics? If so, then perhaps science can again serve humanity at a basic level. But if there is no scientific foundation for ethics, then how can ethics influence science—let alone science's exuberant but wanton godchild, technology. It boils down to the classical-machine argument:

If our actions are determined by forces beyond our control, it seems futile to invoke ethics or values to guide them.

Some authors suggest that the crisis in values will be resolved if students go back to reading classics, such as Plato, but I submit that the malaise goes deeper.[2] Our science has progressively discredited religious prejudice and rigid dogma and has undermined the practice of primitive rituals and engagement with mythic lifestyles, but it has also compromised what is abiding in religious teachings, rituals, and myths—values and ethics. Can we restore values and ethics free of dogma? Can we understand values and ethics stripped of their mythological base?

Perhaps not, but the chances improve if science itself can establish that ethics is part of the universal scheme of things. Without a scientific foundation, ethics continues to be expressed in a culture-bound and arbitrary way. As an example, consider scientific humanism, which promotes human values. Humanists say, do unto others as you would have others do unto you, because if you don't, you won't be accepted in the human community. But this formula does not work. It is a reactive stance, and ethics is fundamentally proactive.

Any arbitrary standard is clearly antithetical to science. Similarly, the recent talk about the establishment of ethical standards in the practice of science and technology is hollow unless ethics can be established on firm scientific principles. It seems essential to recognize *the establishment of ethics and values as a genuine scientific pursuit.*

Recent developments in quantum physics already suggest the possibility of a fundamental contribution from physics to the subject of ethics and values. Alain Aspect's experiment strongly indicates that our separateness from the world is an illusion. From these data alone, some people take heart that the quantum worldview permits, and even demands, ethics and values.[3]

With the idealist interpretation of quantum mechanics, we can go even further. Once we can understand the conditioned camouflage that clouds the tangled-hierarchical mechanism of our brain-mind and creates the illusion of ego-separateness, it is only one more step to develop a science of ethics that will enable us to live in harmony with the scientifically established principle of inseparability. In developing this program, our spiritual/religious heritage can be very helpful. A bridge between the scientific and spiritual philosophies of idealism will genuinely heal the divisions in society that challenge and too often compromise ethics and values.

The basic principles of such a science are already clear. Ethics must reflect our quest for happiness, which lies in resolving internal value conflicts. In other words, ethics must be a guide to a movement toward wholeness—a guide to the integration of our classical and quantum selves. Another basic principle is the fundamental inseparability of ethics and creativity. The new ethics cannot be calcified by ritual belief systems. Instead, it must flow meaningfully from the human being's pursuit of inner creativity. Clearly such ethics must sometimes contradict material-realist beliefs.

As such a science develops, we will be able, at the most personal level, to take responsibility for the world that is us. As Viktor Frankl once commented, we must supplement the Statue of Liberty on the East Coast with the Statue of Responsibility on the West Coast. This will mean large numbers of us living a life rich in inner creativity. In such a world, we may even approach the elusive goal of peace within, as well as between, each of us.

Before we consider details of a new science of ethics, let us review the two systems of ethics that have dominated Western thought.

THE KANTIAN CATEGORICAL IMPERATIVE

According to the eighteenth-century German philosopher Immanuel Kant, the question of morality is one of individual motive. Kant believed that motivation comes from an idea domain and that all human beings have an intuitive sense of what their moral duty is in a general way. Thus a categorical imperative hangs over us to perform these duties. Why should I be moral? According to Kant, we hear an inner imperative: Do your duty. This imperative is the inner moral law that each of us legislates for ourselves. Morality consists of performing these duties, regardless of inclination or disinclination. In addition, Kant suspected that these duties are universal laws. They apply to all human beings in a rational, harmonious way such that conflicts between one person's duty and another's do not arise.[4]

What are these duties? Kant believed that they are founded in rationality and that, using reason, we can discover them. We can do so by asking ourselves, Would I want this action that I am contemplating to be universal? If such a thing is desirable, then we have discovered a universal law. There is more than a little circularity in this argument.

The Kantian theory of ethics is an interesting blend of idealist and realist aspects. He poses an idea domain wherefrom arise the categorical imperatives. This is clearly an idealist metaphysic. We apply the moral law to ourselves, make a decision, and take responsibility for it. This is clearly in consonance with idealist views. Kant also seems to believe in objective universal moral law—a realist belief. It is in this regard that Kant goes astray. (Certainly, the universality of Kant's moral law is questionable if only from the empirical observation of genuinely ambiguous situations that challenge our knowledge of right and wrong with the utmost subtlety.)

Kant also correctly perceived that the inner moral law is an intimation from a free, immortal soul. Unfortunately, he believed we were without access to this inner self.

For Kant, where ethics ends, religion begins—along with its system of reward and punishment. Simplistically, religions maintain that we are rewarded for good deeds with an afterlife in heaven or punished for our misdeeds with an afterlife in hell.

The Material-Realist Position: Utilitarianism

Utilitarianism, often summarized by the dictum "the greatest happiness for the greatest number," was developed by the philosophers Jeremy Bentham and John Stuart Mill in the nineteenth century.[5] It continues to dominate the Western psyche—especially in the United States. Happiness is defined basically by pleasure: Whatever brings about the greatest amount of pleasure to the greatest number of people is the ultimate good.

Utilitarianism is an interesting mixture of materialism, locality, objectivity, epiphenomenalism, and determinism—all the elements of material realism. Only material (objective and local) things, such as gold, sex, power—the objects of hedonism—bring happiness. So we must pursue them. Lest this seem like the philosophy of hedonism, sprinkle in a little socialism whereby individual happiness is not the issue. It is the happiness of the society on the average that we should maximize. A war will inflict pain on a few, but it is justified if it will bring happiness to the majority.

According to utilitarianism, ethical considerations are objective. By studying an action's consequences for pleasure or pain, we can assign to it a happiness value and an unhappiness value vis-à-vis the whole society. Bentham even developed an absurd hedonic calculus to calculate the happiness index of an action.

Many philosophers concede that even under utilitarianism we must be free to pursue the right course. Upon closer examination, however, we see that lurking behind this philosophy is a strict belief that subjectivity (or personal choice) in a moral issue is an irrelevant epiphenomenon and does not play any ultimate role. That is, we may think that we are choosing, but that is illusory thinking. Events and actions follow a natural (deterministic) law. Ethical theory enables us to predict the outcome and thereby gain control (by siding with the so-called good). Neither does intuition about the goodness or badness of an act play any role, because in this philosophy intuition does not exist.

Finally, utilitarianism makes no mention of personal responsibility: We are creatures of determinism. So long as ethical considerations follow an objective science of ethics (the realist science of ethics), all is consonant with the philosophy of determinism: The questions of choice and responsibility do not arise.

Even today, however—when, at the societal level, we seem to make most ethical decisions based on the utilitarian philosophy—at the personal level, Kant's thinking still touches us. Many people still follow the moral law within or are tormented by it—or both. Some of us question the validity of such enterprises as the hedonic calculus; others have difficulty with the utilitarian-ethics' aspect of natural law. Many are disturbed by the lack of a place for moral responsibility in utilitarian ethical philosophy.

A growing consensus seems to be that the realist science of ethics in the form of utilitarianism is simply incomplete. It denies the validity or usefulness of many genuine subjective experiences.

IDEALIST ETHICS

Suppose we are not classical machines. What if, as this book claims, we are consciousness manifesting as dual quantum-classical systems? Can we make a more authentic and complete science of ethics in a quantum universe? As soon as we realize that we have the inherent privilege to act in the quantum modality, with freedom and creativity, then the whole argument for subjective aspects of ethics takes on the immediacy of reality. To acknowledge that we are free in our actions is to acknowledge that we are responsible for our actions. Is this, then, the purpose of ethics and values—to be codes of responsibilities, codes of duties, codes of shoulds and shouldn'ts? Quantum theory defines our consciousness as the chooser. Is the purpose of idealist ethics then to define good choices as opposed to bad choices, to categorize right and wrong better than does the realist ethics?

It seems simple at first. For example, consider the golden rule: Do unto others as you would have them do unto you. Can we derive this rule from the idealist metaphysics? Of course, by definition, that is the origin of the maxim: for we are all one consciousness: hurting another is to hurt one's self and vice versa. Loving another is loving one's self.

What if the golden rule is your criterion for making choices, your code of duty? Suppose that you and your best friend have gone boating in a large lake without life jackets. What do you do when the boat sinks? You are not a strong swimmer, but you think you can make it to shore. Your friend, however, does not swim at all and is panicking. If you love yourself, you will wish to save yourself. If you love your friend as yourself, you will attempt to save her. Rationally, the impulse is to take your best chance of survival, yet we know that many times people try to save the other, even when that person is a stranger. Does the golden rule help one solve this dilemma?

The goal of ethics is rightness, goodness. It is to this end that we conscientiously learn ethical rules, such as the Ten Commandments or Buddha's Eightfold Way—codes developed by illustrious idealist thinkers. We naively assume that if we memorize the rules, they will pave for us a well-charted road with clearly marked intersections, a road that will take us safely through the vicissitudes of life to that pinnacle where we will be clearly revealed as a Good Person, an Ethical Person.

Alas, it is not so simple, as we all discover rudely enough. We discover the difference between the letter of the law and its spirit. We discover that there can be conflict between interpretations or versions of the good, as in the above boating-accident scenario. We discover that rewards and punishments are not justly distributed on the basis of ethical merit. Pranksters have destroyed or misaligned the signs at many important intersections along our Road to the Pinnacle of Good. This is why many books on ethics, written by wise and thoughtful people, have not really solved the problem of ethics for us. In a beautiful case analysis of an ethical conflict, Sartre concluded that ultimately people must choose the path according to their instinct or feeling.[6] What is Sartre talking about?

We can examine Sartre's thought by applying the ideas of classical and quantum modalities from the quantum theory of the self. First of all, both modalities are active in us. Although we have freedom of choice in the quantum modality, we are also classically conditioned beings with a tendency to respond as if we were classical machines. This tendency to avoid choice extends to a tendency to avoid responsibility. We want to be free in the quantum modality, and yet we want to have a chart for that freedom. Unfortunately, any path that is charted is a classical path—a fixed path—and does not necessarily lead straightforwardly to an ethical destination in all situations.

This essential predicament has to be understood. Sartre understood it, and this is what existential ethics is all about. Understanding the difficulty of applying general ethical principles to infinitely varied specific circumstances helps us to accept some inconsistencies in the ethical behavior of ourselves and others. It helps us to become less judgmental.

So ethics is impossible to formulate without talking about manifesting ethics in life. Interestingly, this also serves to answer Kant's (and everybody's) question, Why am I moral?

WHY AM I MORAL?

It is ironic that ethical principles have been passed down faithfully through the generations of humanity without equally meticulous instructions for how to manifest ethics. Without an explicit context of devotion to growth toward transformation, it simply is not possible for a person truly to live by these principles. Properly understood, ethical codes are not primarily rules for external behavior but instructions for internal meditation as we behave externally. They are techniques to manifest freedom in us, to facilitate our ability to act in the quantum modality. Thus the maxim Love thy neighbor as thyself is useless to most of us as a code of behavior because we do not truly love ourselves and, therefore, do not really know what love is in the first place.

At the heart of this injunction lies the realization that we are not separate from our neighbor. Therefore, loving one's self is loving one's neighbor and vice versa. So the assignment is simply, learn to love. Love is not a thing but an act of being. Love as a meditation practiced as continuously as possible is different from love as a set of prescribed behaviors or as a pleasure response. Love as a meditation allows us to soften our ego-boundaries a little—to permit our neighbor's consciousness into our awareness once in a while. With patience and perseverance, love does happen within us. And that love—not externally imposed or derived forms of behavioral love—is what transforms our behavior and touches our neighbor.

So here is the answer to a question that inevitably arises when we study Kant's ethical philosophy. If "Do your duty" is a universal categorical imperative, why are only some of us tormented by it and not others? The answer is, first, as Kant himself recognized, ethics and the inner moral laws are intimations from our inner self to know our full self. Second, and more importantly, the injunction to do your duty touches only those of us who are committed to explore our full self, to awaken to the buddhi level beyond ego. If we are mired down in our ego-identity, we gradually lose the ability to hear these inner commands.

It is interesting that religions touch the right chord with their idea of reward and punishment. The reward for moral action is indeed heaven, but not in the afterlife. Heaven is in this life; it is not a place but an experience of living in quantum nonlocality. Similarly, to avoid the ethical imperative is to perpetuate the ego-level existence and to doom one's self to a living hell.

What is sin? It is important to ask this question because organized religion often focuses its energy and influence on ideas of sin, of good versus evil, of reward and punishment. Most organized religions envision some version of hell as punishment for sin after death. Most also provide for forgiveness or absolution of sin before death to enable the sinner to escape hell.

In a quantum view of ethics, the only sin is that of completely fossilizing the self or others in classical functioning, to block one's own or another's access to the quantum modality and to the manifestation of freedom and creativity. (This is entirely consistent with the Christian idea of original sin as the separation from God.) For condoning this stasis, we do end up in hell—the hell-on-earth of ego-bondage, as suggested in the following story:

A good man died and, as expected, found himself in a heavenly place. He was hungry, so he asked an attendant for food.

"All you have to do to get food is to desire it," he was told.

Wonderful! But after he had eaten his custom-desired feast, he was lonely. "I want some female companionship," he told the attendant, and once more he was told that he need only desire it. So he desired and again was content for a while with his beautiful companion.

Then he began to feel bored and once more approached the attendant. "This isn't what I expected," he complained. "I thought one was bored and dissatisfied only in hell."

The attendant looked at him and asked, "Where do you think you are?"

Our ego-selves too often attempt to find a balance by averaging polarized concepts, such as good and evil. This dualizing tendency of the classical modality causes a lot of trouble because it leads, whether intentionally or not, to judgment by absolute standards. Such judgments as often as not limit a person's potential. They certainly limit the judger's potential and frequently also limit that of the judged. It is not our moral prerogative to enforce a code of ethics—or any code—on another, because doing so interferes with the other's freedom. (This is not to say that we cannot restrain a person who overtly and unmistakably is threatening the freedom of others. Social utilitarianism has its place within idealist ethics—just as scientific realism has its place within monistic idealism.) Imagine how many conflicts would disappear from the world if no one ever imposed an ideology on another!

The transformative good that we seek is that of the quantum modality—the good that transcends the polarities of both good and evil. It is the good of the atman consciousness.

Preaching what is not practiced can be dangerous. Most of us can conjure up ugly images of moral rectitude, for history has recorded unspeakable cruelty in the name of morality. Gandhi understood the cardinal rule of ethics: Ethics must be a spiritual practice with pure inner roots. A woman once brought her small daughter to Gandhi with a simple request: "Tell my daughter not to eat candy. It is bad for her teeth. She respects you, and she will obey you."

But Gandhi refused. "Come back in three weeks," Gandhi told her. "I will see what I can do."

When the woman returned in three weeks with her daughter, Gandhi took the little girl on his knee and gently instructed her, "Don't eat candy. It is bad for your teeth."

The girl shyly nodded her assent. Then she and her mother left for their home. When they were gone, some of Gandhi's associates were upset and confronted him. "Bapu, did you know that woman and her child had to walk for hours to see you, and you made them walk that distance twice in three weeks? Why didn't you give that simple advice to the little girl when they first came?"

Gandhi laughed. "Three weeks ago I did not know if I could stop eating candy. How could I advocate a value unless I myself can practice it?"

If ethics were a fixed and rational system of behavior, how could it ever be detailed enough to cover all situations and premises in a changing world? Instead, ethical or moral choice is best expressed in an ambiguous manner. Ambiguity breeds creativity, and creativity often is essential to find ethical solutions to dilemmas. For example, let us reconsider the boating-accident scenario previously described. The problem in applying the golden rule in this predicament is that if you were drowning, of course you would want your friend to save you, but if you knew that the attempt would only cost his life in addition to yours, you would want him to save himself. The uncertainty of the situation creates an ambiguity—an inescapable doubt about what is ethical—that only a creative response can resolve.

The Russian physicist Yuri Orlov, whose recent theory of doubt was developed in a prison cell, sees in the development of healthy doubt the characteristic of a double-bind. The informational input creates two competing situations in the mind of the doubter, who cannot withdraw. The resolution, according to Orlov, is not a flip of a coin, but creativity: "It is essential that there exist a conflict: On the one side it is impossible to solve the dilemma, on the other side it is necessary to solve it—and rely on one's own inner voice, and not say a generator of random numbers."[7]

According to Orlov, doubt occurs because there is no logical solution. Logic yields only a paradoxical oscillation between the options. The same is true for a moral dilemma. When logic is insufficient to reach an ethical answer, such an answer can only be reached by a creative quantum leap. Even when logic can be stretched to reach a parsimonious solution, a creative approach often yields a richer solution that actually revolutionizes the context of the problem. Ethics in its essence seems to involve inner creativity, a transformative encounter with our quantum self. This is the implicit message in the turn-the-other-cheek forgiveness preached by Jesus, which is so hard for us to accommodate in our classical modality.

It is this access to the quantum buddhi self that we idealize but find so hard to live by in our responses to personal affronts. To maximize access to the quantum self, to maximize creativity and free will, we must be committed to radical transformation of the psyche. It is a fantasy to expect otherwise. The mistake made by most prophets has been their lack of emphasis on the transformational motivation as fundamental. Externally applied prescriptions are strictly Band-aid therapy. No, people ordinarily are not capable of manifesting an ideal without getting into seemingly insurmountable conflicts with conventional ideas of fair play, reward and punishment, and other social agreements that support the pursuit of happiness and the so-called good life.

In the quantum modality, we avoid preconceived answers: Creativity is the goal; we must remain open to more expansive possibilities without automatically, as an act of classical conditioning, taking the shortcut of a pre-given ethical formula. Empowering people to find, for example, miraculous solutions in situations like that of the drowning friends in the lake is the goal. Such creative intervention is surely involved when a middle-aged woman lifts a truck off an injured son or husband. It is in ethics that we experience perhaps our greatest potential for freedom.

Thus we can define the fundamental idealist ethical principle as the preservation and enhancement of our own and others' access to the quantum modality—to the buddhi level of being (which includes both freedom and creativity).[8] Let us now analyze the step-by-step approach (the different stages of spiritual life) laid out in the idealist literature from the point of view of an ethical journey of manifesting morality in our life. For the journey of inner creativity is not complete until the product, the transformation of our self, is complete and is communicated for others to see.

THREE STAGES OF IDEALIST ETHICAL PRACTICE

One of the best expositions of the idealist literature is the Bhagavad Gita, which we will follow in this summary. According to this thinking, the human ethical journey is seen in terms of three spiritual paths—the yoga of action (karma yoga), the yoga of love (bhakti yoga), and the yoga of wisdom (jnana yoga). In each stage of human ethical development beyond the utilitarianism of the ego, one of these yogas dominates—although all three yogas are practiced simultaneously. Each of these yogas contains a practice of ethical action.

In the first stage, the yoga of action, one practices how to act without attachment to the fruit of the action. It is the ego's coveting fruit from the action that interferes with seeing clearly the nature of our conditioning. This inability to see our conditioning prevents us from recognizing our duties and keeps us from ethical actions. This is the preparation stage. We begin to see our conditioned actions so that we can choose to act morally. This stage sometimes culminates with the realization of our fundamental oneness with the world—the ah-ha experience of inner creativity.

At the next stage, the yoga of love, we act in service to others (as God's instrument, to use a more religious metaphor). This is the altruistic stage, the central stage of ethical and moral action. We discover otherness—the independent rather than contingent validity of other individual manifestations. We hear the call of duties and heed it. We serve in direct and immediate ways the good of all, not just the abstract greatest good for the greatest number. We do not compromise fundamental moral duties once we see what those duties are. Our service opens our heart to love others. The more we love, the more we are able to act ethically toward ourselves and others.

In the third stage, the yoga of wisdom, we act through a perfect alignment of our will and the will of the quantum modality of the self. In this alignment we surrender the ego-level will to the moment-to-moment choice of the unitive consciousness. This is similar to the Christian ethical doctrine, Thy will be done. However, the latter way of putting it can lead to a major misconception if "thy" is interpreted as being separate from "I." This separation suggests giving up one's free will to some external agency, but "thou" is not separate from "I" when one arrives at this stage of maturity. Thus in surrendering the ego to the quantum modality, one becomes truly free and creative. Strictly speaking, ethics and morality are no longer needed as guides because there no longer is any conflict. All these—ethics, morality, conflicts—dissolve into the will of the unitive consciousness. Then there is only appropriate action.

Finally, let us consider a question that bothers a lot of ethical philosophers. What if the moral life conflicts with the so-called good life? This, of course, depends on how one defines the good life. As we transform from the ego level to the buddhi level of being, the definition of the good life as the pursuit of happiness gradually changes to that of a life of joy. Continual pursuit of transitory pleasures yields to a stable, easy-without-effort living in wholeness, but the moral life is a life of service. Can the two conflict? The practicing idealist discovers, as did the poet Rabindranath Tagore,

> *I slept and dreamt that life was joy.*
> *I awoke and saw that life was service.*
> *I acted and behold, service was joy.*

Chapter 19

SPIRITUAL JOY

You have, in this book, the basic idealist schema of self-exploration beyond ego. Is it religion or is it science? And what is the role of philosophy?

Religion comes from the root word *religiere* meaning "to connect again." The culmination of the adult developmental process is, indeed, in reconnecting with what we originally are—with the primary processes of our brain-mind, with the nonindividual self. So the idealist program is indeed a religion in this sense.

However, in all major religions there are dualistic tendencies. In most religions there is a deification of a particular teacher or promulgation of a particular teaching or belief system. These have to be transcended in the final reckoning. So, in its last developmental stage, the idealist schema must go beyond all religions, creeds, belief systems, and teachers.

Is the idealist schema a science? I believe that most if not all of the stages of adult development can be put to objective tests (in the sense of weak objectivity) and thus can qualify as science. On the psychology of liberation we have nothing, said the psychologist Gordon Allport not so long ago. Well, here, finally, is a psychology of liberation.

When we look at the phenomenon of the human spiritual quest as the newest extension of psychology, perhaps the final rapprochement between science and religion will be accomplished. In this psychology of liberation, science and religion will have complementary functions. Science will be involved in further objective studies, both theoretical and practical, relating to the phenomenon. Religion will be involved in dissemination of the scientific knowledge thus gained, but in a subjective way because the objective teaching of such knowledge is largely irrelevant. Crowning both and acting as the signpost for both will be philosophy—the idealist metaphysics, which will continue to be enriched by new insights.

The fundamental unverifiable (in the scientific sense) idealist metaphysic is a one-liner: Consciousness is the ground of all being and our self-consciousness is That consciousness. But the simplicity of this dictum is also its richness. Witness the vast philosophical literature in which people have tried to expound and explain this metaphysic at various times and in various cultures. This book is the latest contribution to the ongoing idealist enterprise—a contribution appropriate for our predominantly scientific culture.

Within the spiritual traditions two significant proposals as to the spiritual way of life have surfaced: The dominant one is based on world-negation. The phenomenal world is dukkha—unease, suffering—said the Buddha. In Pauline Christianity, the entire life of a Christian is a penance for the original sin. In much of Hindu Vedanta philosophy, the phenomenal world is seen as illusion. People in this tradition have emphasized enlightenment, renunciation, nirvana, salvation as various stages and forms of escape from the illusory world of suffering. We turn to the spirit because the material world has nothing to offer us; we declare spiritual upliftments to be the highest virtues. From this position, science, which is the exploration of the world, seems oppositional and contrary to spiritual traditions, and this apparent dichotomy spawned antagonism between science and spirituality.

Within the spiritual disciplines, however, there have always been, although never dominant, insistent world-affirming voices. Thus in Japan, alongside Rinzai Zen with its emphasis on enlightenment, there has also been Soto Zen, which emphasizes the awakening of compassion so that we can serve the world. In India, among all the world-negating Upanishads, one, the Isha Upanishad, stands out by its declaration of enjoying immortality in life itself.[1] In China, the Taoists similarly have proclaimed a philosophy of peace and joyful living in the world. The Bauls of India, too, have sung the glory of spiritual joy.

Because of its world-affirming character, spiritual joy welcomes the exploration of manifest nature, which is the primary activity of conventional science. So it is not surprising that, finally, we have developed a science—idealist science—that is truly integrated within a spiritual philosophy of joy. This idealist science challenges the world's religions to change their emphasis, to recognize both fundamental joy and suffering, both the world and the spirit. Realization of this goal will be the final rapprochement between science and religion.

Beyond science, religion, and philosophy, there exist us and our free will. In one of the last verses in the Bhagavad Gita, Krishna tells Arjuna to make a decision out of his own free will whether to live in the idealist manner or not. This is the decision that you, I, and all of us have to make out of our own free will.

In poll after poll, it has been revealed that an amazingly high percentage of Americans have had mystical experiences. If only they would make those experiences the basis for awakening to the buddhi level of being! And when a significant number of us become thus re-enchanted, being and living in the buddhi, a change in the movement of consciousness may very well occur throughout the world.

I believe that such a massive movement of consciousness can be called a renaissance. Such transitional periods have occurred in many cultures and civilizations. The next such renaissance, which perhaps is a-birthing, will be a very special one since, thanks to modern communications technology, humanity is now interconnected. The next renaissance will have worldwide reverberations; it will be a global renaissance of peace.

The Bhagavad Gita portrays such events of renaissance as the coming of an *avatara*, or world-teacher. In the past such avataras have sometimes been isolated, single individuals; at other times there have been collections of individuals. But the world is much bigger now and needs an unprecedented number of individuals to become avataras to lead the next renaissance. Imagine your journey and mine toward a time when there is a vast uplifting of humanity from fragmentation into unity in diversity. That would be truly a hero's journey.

Hero's Journey

Myths in many cultures include a theme to which the mythologist Joseph Campbell refers as the hero's journey.[2] The hero suffers a separation from his world, sets out on his own to confront mysterious forces, and finally returns in glory, carrying with him (to a magnificent reunion) the knowledge that he has gained. The Greeks expressed their appreciation for the benefits of fire in the Prometheus myth: Prometheus went to heaven, stole the secret of fire from the gods, and presented it as a gift to mankind. In India, Gautama the Buddha renounced the comforts of his princely world to go through the hero's journey that led to his nirvana. He returned thence to teach the truths of the Eightfold Way. Moses, the hero of Israel, sought his God on Mount Sinai, received the Ten Commandments, and returned with them to unite his people. In each case, the reunion brought forth a teaching of integration—a new way to manifest the spirit in the experience of ordinary life.

I see the myth of the hero's journey being played out again in science's search for the nature of reality. The individual heroism of the old days has given way, however, to collective heroism. Many unheralded scientists have trod the heroic path through each of the three stages of the myth.

The Cartesian separation of mind and matter was historically unavoidable in order for science to pursue a free course unshackled by theology. It was necessary to study unconscious matter without theological bias in order to gain an understanding of the mechanics and interactions that shaped all matter, including the living and the conscious. It took almost four hundred years to achieve the relative mastery we now enjoy over these physical forces.

There have been many landmarks on this journey of separation, and many heroes. Descartes set sail, and very quickly Galileo, Kepler, and Newton became the helmsmen of the heroes' ship. Darwin and Freud completed the separation by extending the laws of mechanics to the arena of the living and the conscious, and the separation has been maintained by hundreds of scientist-sailors.

In the twentieth century the wind freshened in a new direction for the heroes' ship. Planck discovered the quantum, Heisenberg and Schrödinger discovered quantum mechanics, and together these discoveries forever altered the old materialist, separatist course. As Bertrand Russell put it, in twentieth-century science matter looked less material and the mind less mental. The four-hundred-year gap between the two was ready to be bridged: The return of the hero had begun.

Prometheus brought back fire. Buddha brought back the Eightfold Way. Each return resulted in a revolution in the dynamics of a society, in a full-blown paradigm shift. Today, in quantum mechanics and its interpretation and assimilation within an idealist science, we see the paradigm-changing potency of Prometheus' fire and Buddha's noble truths.

Mythology is the history of the play of consciousness. If you refuse to explore consciousness, if you fail to reject the idea of consciousness as an epiphenomenon, then the myth may pass you by. The climax, the hero's return, of the most versatile of all myths is now in play, but few can see it clearly. Such blindness prompted the author Marilyn Ferguson to dub the developing paradigm shift the Aquarian conspiracy, but it is the most open conspiracy that history has ever recorded.[3]

The legacy of the old separatists—mind-body and matter-consciousness dualism—will not go away by asserting a monism based on material realism, as many scientists of the mind tend to do. As the Canadian neurosurgeon Wilder Penfield emphasized, "To declare [that] these two things [mind and body] are one does not make them so." Indeed it does not. New schisms simply replace the old when a monistic view is hastily embraced—one that is inconsistent and that does not heed the legitimate concerns of idealists (that is, how to include body, mind, and consciousness, all three elements, in our model of reality).

The paradigm outlined here considers truly integrated ideas that take into account the concerns of both the idealist and the materialist camps. These ideas are being considered not only in the theories of quantum physics but also in experimental laboratory work in cognitive psychology and neurophysiology.

Much work remains to be done. Even though the new view gives a consistent interpretation of quantum mechanics and resolves the mind-body paradoxes, a host of questions must be answered before a coherent picture emerges. If consciousness is the fabric of the world, how do we find new laboratory experiments to confirm the idea? This is but one of the remaining unanswered questions.

The ideas explored here of a new consciousness-based idealist science—ideas growing out of the efforts to integrate science into the philosophy of idealism—are worth your serious, personal appraisal. If that assessment leads you to explore consciousness, to begin your own hero's journey of transformation, my work will have been justified.

For hundreds of years we have bowed to the objectivity of science but have cherished subjectivity and religion in our living. We have allowed our lives to become a set of dichotomies. Can we now invite science to help integrate our ways of living and revolutionize our religions? Can we insist that our subjective experiences and spiritual philosophy be allowed to extend our science?

"Someday," said the Jesuit philosopher Teilhard de Chardin, "after we have mastered the winds, the waves, the tides, and gravity, we shall harness . . . the energies of love. Then for the second time in the history of the world man will have discovered fire." We have mastered the winds, the waves, the tides, and gravity (well, almost). Can we begin harnessing the energies of love? Can we realize our full potential—an integrated access to our quantum and classical selves? Can we let our lives become expressions of the eternal surprise of the infinite Being? We can.

GLOSSARY

Amplitude: The maximum change of a wave disturbance from the equilibrium position.

Anthropic principle: The assertion that observers are necessary to bring the universe into manifestation; also called the *strong anthropic principle.*

Archetype: A Platonic idea that is the precursor of a material or mental manifestation; also the Jungian symbol of the instincts and primordial psychic processes of the collective unconscious.

Aspect, Alain: The experimental physicist at the University of Paris-Sud acclaimed for the 1982 experiment named after him that established quantum nonlocality.

Atman: The Sanskrit word meaning higher cosmic self beyond ego, adapted in this book as the term for the quantum creative self.

Awareness: The "space" of the mind in relation to which objects of consciousness, such as thoughts, can be distinguished; analogous to physical space in which material objects move.

Behaviorism: The primary paradigm of psychology in the twentieth century; it holds that the explanation of human behavior is to be found in the history of stimulus-response-reinforcement patterns of a person.

Bell inequalities: A set of mathematical relationships between possible results of observation of correlated quantum objects, derived by John Bell on the assumption of the locality of hidden variables.

Bell's theorem: A theorem discovered by Bell asserting that local hidden variables are incompatible with quantum mechanics.

Bhakti yoga: The yoga of love or devotion.

Binary message: A message using variables that take on one of two possible values, 0 or 1.

Blindsight: Sight without conscious awareness of it.

Bohm, David: An English physicist who has contributed substantially to the problem of interpretation of quantum mechanics. Although a realist, Bohm greatly appreciates the transcendent domain.

Bohr, Niels: A Danish physicist, discoverer of the Bohr atom and of the complementarity principle. During his lifetime, he was the most influential spokesperson of the Copenhagen interpretation. According to Heisenberg, he never subscribed to the positivistic philosophy (and instrumentalism) that later became the hallmark of many physicists' understanding of quantum mechanics. Bohr completely understood that there was significance in the strangeness of quantum physics.

Brownian motion: The random motion of particles suspended in a liquid, caused by random collisions of the particles with the molecules of the liquid.

Categorical imperative: Philosopher Immanuel Kant's idea that we act morally because we hear inner injunctions to do our moral duties.

Causal determinism: See *Determinism.*

Causality: The principle that a cause precedes every effect.

Cerebral cortex: The outermost and most recently evolved segment of the mammalian brain; also called the neocortex.

Chaos theory: A theory of certain deterministic classical systems (called chaotic systems) whose motion is so sensitive to initial conditions as not to be susceptible to long-term predictability. To materialists, this determined but not predictable character of chaotic systems makes them an apt metaphor for subjective phenomena.

Circularity: See *Self-reference.*

Classical functionalism: See *Functionalism.*

Classical mechanics: The system of physics based on Isaac Newton's laws of motion; today it remains only approximately valid for most macro objects as a special case of quantum mechanics.

Classical self: A term used in this book to denote the conditioned modality of the self, the ego.

Coherent superposition: A multifaceted quantum state with phase relations among its different facets (or possibilities). For example, an electron going through a double slit becomes a coherent superposition of two states: one state corresponding to its passing through slit 1 and another state corresponding to its passing through slit 2.

Collective unconscious: Unitive unconscious—that aspect of our consciousness that transcends space, time, and culture, but of which we are not aware. A concept first introduced by Jung.

Complementarity: The characteristic of quantum objects possessing opposite aspects, such as waveness and particleness, only one of which we can see with a given experimental arrangement. According to the present author, the complementary aspects of a quantum object refer to transcendent waves and immanent particles.

Consciousness: The ground of being (original, self-contained, and constitutive of all things) that manifests as the subject that chooses, and experiences what it chooses, as it self-referentially collapses the quantum wave function in the presence of brain-mind awareness.

Copenhagen interpretation: The standard interpretation of quantum mechanics, developed by Bohr and Heisenberg, that is based on the ideas of the probability interpretation and the principles of uncertainty, complementarity, correspondence, and the inseparability of the quantum system and its measuring apparatus.

Correspondence principle: The idea, discovered by Bohr, that under certain limiting conditions (which are satisfied by most macrobodies under ordinary circumstances) quantum mathematics predicts the same motion as Newtonian classical mathematics.

Creativity: The discovery of something new in a new context.

Decay: The process in which an atomic nucleus emits harmful radiations and transforms to a different state.

Democritus: The ancient Greek philosopher known primarily for founding the philosophy of materialism in the West.

Determinism: The philosophy according to which the world is causal and completely determined by Newton's laws of motion and initial conditions (the initial positions and velocities of the objects of the space-time universe).

Diffraction pattern: A pattern of alternate reinforcement and cancellation of wave disturbances produced whenever waves bend around obstacles or pass through slits.

Distant viewing: Viewing at a distance through psychic telepathy; in the model of this book, nonlocal viewing

Double-slit experiment: The classic experiment for determining characteristics of waves; a wave of light, for example, is split by passing it through two slits in a screen to make an interference pattern on a photographic plate or a fluorescent screen.

Dualism: The idea that mind (including consciousness) and brain belong to two separate realms of reality. This philosophy, however, cannot explain how the two realms interact without contradicting the conservation of energy that we see in our world.

Ego: The conditioned aspect of the self.

Eightfold Way: The eight principles of living enunciated by the Buddha for the cessation of the fundamental unease (*dukha*) of the human condition.

Einstein, Albert: Perhaps the most famous physicist that ever lived, he is the discoverer of the relativity theories. He was a major contributor to quantum theory, including the basic ideas of wave-particle duality and probability. In his later years, he found the instrumentalist (and positivistic) trend of the interpretation of quantum physics distasteful to his scientific beliefs.

Epiphenomenalism: The idea that mental phenomena and consciousness itself are secondary phenomena of matter and are reducible to material interactions of some substructure.

Epiphenomenon: A secondary phenomenon; something that exists contingent on the prior existence of something else.

Epistemology: The branch of philosophy that studies the methods, origin, nature, and limits of knowledge; it is also the branch of science that studies how we know.

EPR correlation: A phase relationship that persists even at a distance between two quantum objects which have interacted for a period and then stopped interacting. In the model of this book, the EPR correlation corresponds to a potential nonlocal influence between the objects.

EPR paradox: A paradox invented by Einstein, Podolsky, and Rosen to establish the incompleteness of quantum mechanics; instead, the paradox paved the way for the experimental proof of quantum nonlocality. See *EPR correlation.*

Evoked potential: An electrophysiological response produced in the brain by a sensory stimulus.

Faraday cage: A metallic enclosure that blocks all electromagnetic signals.

Feedback system: A hierarchical system in which the lower level affects the upper level, and the upper level reacts back and affects the lower. An example is a thermostatically controlled room.

Free will: Freedom of choice undetermined by any necessary cause. According to this book, at the secondary level we exert free will when we say no to learned, conditioned responses.

Frequency: The number of wave cycles per second.

Freud, Sigmund: The founder of modern psychology, he is an enigma to those who classify people in rigid philosophical categories. Although much of his writing supports material realism, his concept of the unconscious does not fit this philosophy and has been attacked for that reason.

Functionalism: A philosophy of the brain-mind in which the mind is looked upon as the function and the brain as the structure, in parallel to the corresponding computer analog of software and hardware.

Game theory: An idealized study of games, assuming that the players are all rational. In particular, a *zero-sum game* refers to a game in which there is a winner and a loser.

Global workspace: See *Mind field.*

Gödel's theorem: The mathematical theorem that every substantial mathematical system must be either incomplete or inconsistent; there is always a proposition that a mathematical system cannot prove within its own axioms, and yet we can intuit the validity of the proposition.

Ground state: The lowest energy state of quantum systems.

Gunas: Qualities of consciousness in ancient Indian psychology that correspond to psychological drives in more modern terminology. There are three gunas: *sattwa* (creativity), *rajas* (libido), and *tamas* (conditioned ignorance).

Heisenberg, Werner: A German physicist and co-discoverer of quantum mechanics, he was perhaps the only one among the founders of quantum physics to really understand and advocate the idealist nature of quantum metaphysics. His discovery of quantum mechanics is widely regarded as one of the most creative events in the history of physics.

Hidden variables: Unknown (hidden) parameters that are posited by Bohm and others to restore determinism to quantum mechanics; according to Bell's theorem, any hidden variables must reside in a world outside of space-time and, therefore, are inconsistent with material realism.

Hofstadter, Doug: A physicist and researcher of artificial intelligence; he is the author of the book *Gödel, Escher, Bach.*

Holism: The philosophy based on the idea that the whole is functionally or meaningfully more than the sum of the parts.

Homunculus: The "little person" in our head hypothesized as the determinator of our actions.

Idealism: The philosophy that holds that the fundamental elements of reality must include the mind as well as matter. See also *Monistic idealism.*

Identity theory: The philosophy based on the idea that every mental state corresponds to and is identical with a particular physical state of the brain.

Immanent reality: See *World of manifestation.*

Instrumentalism: The philosophy that regards science as just an instrument for analyzing experimental data and guiding new technology and that gives science no credibility whatsoever in metaphysical matters.

Interference: The interaction of two waves incident in the same region of space that produces a net disturbance equal to the algebraic sum of the individual disturbances of the respective waves.

Interference pattern: The pattern of reinforcement of a wave disturbance in some places and cancellation in others that is produced by the superposition of two (or more) waves.

Inviolate level: The transcendent domain beyond the logical discontinuity of a tangled hierarchy from which vantage point the cause of the tangle is clear.

Jnana yoga: The yoga based on using the intellect to transcend the intellect.

Jung, Carl G.: The psychologist who founded a major force of modern psychology that carries his name; he is famous for his concept of the collective unconscious and for his visionary insight that physics and psychology one day should come together.

Kant, Immanuel: The idealist philosopher whose ethical philosophy is based on the idea of categorical imperatives.

Karma yoga: The yoga of action, a yoga in which one acts but surrenders personal interest in the fruit of the action.

Koan: A paradoxical statement or question used in the Zen Buddhist tradition to stimulate the mind to make a discontinuous (quantum) leap in understanding.

Law of conservation of energy: The idea, which has been vindicated in every scientific experiment so far, that the energy of the material universe remains a constant.

Liaison brain: The part of the brain that connects it to the mental order of reality in the dualistic philosophy of Sir John Eccles.

Libido: The Freudian term for the life force, also often used to denote sexual drive.

Locality: The idea that all interactions or communications among objects occur via fields or signals that propagate through space-time obeying the speed-of-light limit.

Logical positivism: A pragmatic philosophy according to which we should keep away from metaphysics and consider only that which we can experience or experiment with.

Logical type: A classification of set theory according to category; for example, a set is of a higher category than are its members.

Macrobodies: Large-scale objects, such as a baseball or a table.

Macrorealism: The philosophy that the world is divided into two kinds of objects, quantum micro objects and classical macro objects.

Marcel, Anthony: The cognitive psychologist who has performed what, from the quantum theoretic point of view, may be a crucial set of word-sense disambiguation experiments.

Maslow, Abraham: The founder of transpersonal psychology, which is based on a monistic idealist framework.

Material realism: A philosophy holding that there is only one material reality, that all things are made of matter (and its correlates, energy and fields), and that consciousness is an epiphenomenon of matter.

Matter waves: Material objects such as electrons and atoms (and even macrobodies) have wavelike properties, according to quantum mechanics. Waves of material objects are called matter waves.

Maya: The perceived separateness of the "I" and the world; also translated as "illusion."

Measurement theory: The theory of how an expanded, multifaceted quantum state reduces or collapses to a single facet upon measurement. According to this author, measurement is accomplished only by conscious observation by an observer with awareness.

Mind: In this book, the organization and functions of the brain at the macro level, including the as-yet-uncharted quantum macrostructure that accounts for the mind's nonlocal characteristics.

Mind field: Awareness where thoughts, feelings, and so forth arise.

Monism: The philosophy that mind and brain belong to the same reality.

Monistic idealism: The philosophy that defines consciousness as the primary reality, as the ground of all being. The objects of a consensus empirical reality are all epiphenomena of consciousness that arise from the modifications of consciousness. There is no self-nature in either the subject or the object of a conscious experience apart from consciousness.

Mystical experience: An experience of consciousness in its primacy beyond ego.

Neo-Copenhagenism: A latter-day instrumentalist revision of the Copenhagen interpretation based on the positivistic ideas that there is nothing beyond our experience, that quantum mechanics is nothing but a set of rules to calculate what we can measure, and that there is no quantum metaphysics.

Neocortex: See *Cerebral cortex.*

Newton, Isaac: The founder of classical mechanics.

Nonlocality: An instantaneous influence or communication without any exchange of signals through space-time; an unbroken wholeness or nonseparability that transcends space-time; see also *Transcendental domain.*

Normal modes: Stable modes of excitation or vibration of a system consisting of several interacting parts.

Nucleus: The heavy core of the atom around which electrons revolve.

Objectivity, strong: A theory or statement about reality that makes no reference whatsoever to subjects or to observer involvement; the idea that separate objects exist independent of the observer; one of the major postulates of the philosophy of realism.

Objectivity, weak: The idea that objects are not independent of the observer but that they must be the same irrespective of who the observer is. The objectivity supported by quantum mechanics is weak objectivity.

Ontology: The study of the essence of being or fundamental reality; metaphysics.

Paradigm shift: A fundamental change in the supertheory or umbrella worldview that governs scientific work at a given time.

Personal unconscious: The Freudian unconscious, the arena of genetically programmed instincts and repressed personal memories that affect our conscious actions through unconscious drives.

Phase relationship: A relationship between the phases (conditions) of motion of objects, especially waves.

Photoelectric effect: The dislodging of electrons from metal when high-frequency light shines on it.

Planck, Max: The discoverer of the idea of the quantum.

Planck's constant: One of the fundamental constants of nature, it defines the scale of the quantum domain; it is because of the smallness of this constant that quantum phenomena are usually confined to the submicroscopic world.

Polarization: The two-valuedness of light, the ability of light to align its axis along or perpendicular to any given direction.

Polarization correlation: Two photons related in phase so that if one is collapsed polarized along a certain axis (as manifested by observation), the other is collapsed polarized along the same axis (as determined by observation) irrespective of the distance between the photons.

Polysemous words: Words with more than one meaning that may seem ambiguous in certain contexts: for example, *palm* (a tree or part of the hand).

Positivism: See *Logical positivism.*

Potentia: The transcendent domain of the probability waves of quantum physics.

Probability wave: The wave of a quantum object; the square of the wave amplitude at a point gives the probability of finding the particle at that point.

Pure mental states: The conditions of the quantum mind, made up of the normal modes of the brain's quantum system, postulated in this book; the Jungian archetypes may be examples.

Quantum: A discrete bundle of energy; the lowest denomination of energy or other physical quantities that can be exchanged.

Quantum functionalism: The philosophy proposed in this book that the functional and structural machinery of the brain-mind consist of both classical and quantum components.

Quantum jump: A discontinuous transition of an electron from one atomic orbit to another without going through the intervening space between orbits.

Quantum mechanics: A physical theory based on the idea of the quantum (a discrete amount) and quantum jumps (a discontinuous transition)—first discovered in connection with atomic objects.

Quantum mind: Mental states arising from the quantum machinery of the brain-mind.

Quantum self: The primary subject modality of the self beyond ego in which resides real freedom, creativity, and nonlocality of the human experience.

Radioactivity: The property of certain chemical elements to spontaneously emit harmful radiation while their atomic nuclei undergo decay. Radioactive decay is governed by quantum probability rules.

Rajas: The Sanskrit word for the tendency toward activeness, akin to libido—a psychological drive of Freudian vintage.

Realism: The philosophy that propounds the existence of an empirical reality independent of observers or subjects. See also *Material realism.*

Reality: All that is the case, including both local and nonlocal, immanent and transcendent; in contrast, the universe of space-time refers to the local, immanent aspect of reality.

Reductionism: The philosophy that phenomena or structures at large can be reduced to and be completely described by their components and their interactions.

Relativity: The theory of special relativity discovered by Einstein in 1905 that changed our concept of time from the Newtonian absolute time to a time existing in relation to motion.

Samadhi: The experience of transcendence of the ego-level identity in which one apprehends the true nature of self and things.

Satori: The Zen term for samadhi.

Sattwa: The Sanskrit word for creativity, one of the psychological drives according to Hindu psychology.

Schrödinger, Erwin: An Austrian physicist, codiscoverer with Heisenberg of quantum mechanics, he was opposed to the probability interpretation for quite some time. Later in life, he embraced some elements of the philosophy of monistic idealism.

Schrödinger's cat: A paradox created by Schrödinger to depict the puzzling consequences of quantum mathematics when interpreted literally and applied to macro systems.

Self: The subject of consciousness.

Self-reference: The logical loop of the self referring to itself; see also *Circularity.*

Set theory: A mathematical theory involving sets that are "a Many that allows itself to be thought of as a One."

Solipsism: The philosophy that only one's own self can be proved to exist.

Speed of light: The speed at which light travels (approximately 300,000 kilometers per second); it is also the highest speed in space-time that nature permits.

State of consciousness: Conditions within consciousness of varying degrees of awareness; examples are waking state, deep sleep, dream sleep, hypnosis, meditative states, and so forth.

Strong objectivity: See *Objectivity, strong.*

Synchronicity: Acausal but meaningful coincidences, a term employed by Jung.

Tamas: A Sanskrit term meaning the tendency toward conditioned action in Hindu psychology.

Tangled hierarchy: A loop between levels of categories, a hierarchy that cannot be causally traced without encountering a discontinuity. An example is the liar's paradox: I am a liar.

Transcendental domain: Pertaining to a realm of reality that is paradoxically both inside and outside of physical space-time. According to this book, the transcendent realm is to be interpreted as being nonlocal—it can influence events in space-time by making possible connections without exchange of signals through space-time. See also *Nonlocality* and *Potentia*.

Transcendental experience: A direct experience of consciousness beyond ego.

Transpersonal psychology: The school of psychology based on the idea that our consciousness extends beyond the conditioned, individual ego to include a unitive and transcendent aspect.

Turing machine: A machine that translates one set of symbols for another. A Turing machine is universal and its functioning, in essence, does not depend on its specific representation.

Ultraviolet: Light of higher frequency than visible light; ultraviolet photons are more energetic than visible photons. Also called *black light*.

Uncertainty principle: The principle that such complementary quantities as momentum and position of a quantum object cannot be measured simultaneously with complete accuracy.

Unconscious: The reality of which there is consciousness but no awareness (according to this book); see also *Personal unconscious* and *Collective unconscious*.

Unconscious perception: Perception without awareness of it; in this book, perception for which there is no collapse of the quantum brain state.

Utilitarianism: The theory that ethics is a code for the "greatest good for the greatest number."

Vedanta: The end or final message of the Hindu Vedas, which appears in the Upanishads, that propounds the philosophy of monistic idealism.

von Neumann, John: A mathematician who was the first to postulate that consciousness collapses the quantum wave function; he also did fundamental work in game theory and the theory of modern computers.

von Neumann chain: The infinite chain of quantum measurement; any measuring apparatus that observes a dichotomous quantum object becomes dichotomous itself; a second apparatus that measures the first becomes dichotomous in its turn, ad infinitum.

Wave function: A mathematical function that represents the wave amplitude of quantum probability waves; it is obtained as a solution of the Schrödinger equation.

Wavelength: The length of a wave cycle: the crest-to-crest distance.

Wavicle: A quantum-mechanical transcendent object that has the complementary aspects of transcendent wave and immanent particle.

Weak objectivity: See *Objectivity, weak*.

Wigner, Eugene: The Nobel laureate physicist who gave us the paradox of Wigner's friend and who also, for a time, supported the idea that consciousness collapses the quantum wave function.

World of manifestation: A monistic idealist's designation of the immanent space-time-matter-motion ordinary world of our experience to distinguish it from a transcendent world of ideas and archetypes; note, however, that both transcendent and immanent worlds exist in consciousness—the first as possibility forms (ideas), the second as the manifest result of a conscious observation.

~~Consciousness - a state of awareness which may or may not be demonstratable by physical responsiveness, cognat or responsive faculties of the body.~~

~~Thoughts - mental processes of consciousness which is capable of receiving, assimilating, processing, analyzing, storing, and recalling sensory + psychic information~~

Consciousness - a state of awareness which may or may not be demonstrable by physical responsiveness

Thought - a process of conciousness capable of receiving, assimilating processing, analyzing, storing, recalling sensary or psychic information.

tai chi
Brock courses
vaccine & computer
clean basement
Chrysler - Checkpt.
Dr Clark
Dr Roberts - cataracts
pen/ink - W/C.
Thornbury - Haliburton

tent strips
safety pins

Consciousness
dietal
Cosmology
universe
species
individual

---

thought ← sensory - experience
perception, acceptance, assimilation - storage - recall

sensory stim
perception
acc reject
assimilation
storage
recall

748-8742

NOTES

CHAPTER 1. THE CHASM AND THE BRIDGE

1. A similar comment was made by the physicist Murray Gell-Mann.
2. This comment is attributed to the neurophysiologist John Eccles.
3. This is a paraphrase of a comment made by the cognitive psychologist Ulric Neisser.
4. This materialist bias now influences most scientists, among them the neurophysiologist Roger Sperry, the physical chemist Ilya Prigogine, and the physicist Carl Sagan, just to name a few.
5. This, for example, is the position of the philosopher Karl Popper.
6. Berman (1984).

CHAPTER 2. THE OLD PHYSICS AND ITS PHILOSOPHICAL LEGACY

1. Maslow (1970).
2. Quoted in Capek (1961).
3. See Gleik (1987).
4. Turing (1964).
5. Penrose (1989), p. 418.
6. Feynman (1982).
7. Jahn (1982).
8. Turing, *op. cit.*
9. For the evidence of discontinuity in creativity, see Goswami (1988).
10. Eccles (1976).

CHAPTER 3. QUANTUM PHYSICS AND THE DEMISE OF MATERIAL REALISM

1. Kuhn (1962).

CHAPTER 4. THE PHILOSOPHY OF MONISTIC IDEALISM

1. Plato (1980).
2. Shankara (1975).
3. Dionysius (1965).
4. Goddard (1970), pp. 32–33.
5. The quotations, shown here as notes 6 through 15, were compiled by Joel Morwood in an unpublished paper.
6. Catherine of Genoa (1979), p. 129.
7. Goddard (1970), p. 514.
8. Arabi (1976), p. 5.
9. Scholem (1954), p. 216.
10. Dowman (1984), p. 159.
11. Colledge and McGinn (1981), p. 203.
12. Monsoor was executed for this statement.
13. Shankara (1975), p. 115.
14. John, 10:30.
15. Goddard (1970), p. 293.
16. Arabi (1980).
17. Nikhilananda (1964), p. 90.
18. I am following William James (1958).
19. See Davies (1983).
20. Heisenberg (1958).
21. Mermin (1985).
22. Aspect, Dalibard, and Roger (1982).
23. Stapp (1977).
24. Heisenberg (1958).

CHAPTER 5. OBJECTS IN TWO PLACES AT ONCE AND EFFECTS THAT PRECEDE THEIR CAUSES

1. Squires (1986).
2. Ramanan (1978).
3. Hellmuth et al. (1986), p. 108.
4. Wheeler (1982).
5. Heisenberg (1930), p. 39.
6. Milne (1926).
7. Blake (1981), p. 108.

CHAPTER 6. THE NINE LIVES OF SCHRÖDINGER'S CAT

1. Lowell (1989).
2. See Gibbins (1987).

3. Everett (1957) (1973). For a good review of the many-worlds theory, also see DeWitt (1970).
4. von Neumann (1955); London and Bauer (1983); Wigner (1962); Wheeler (1983); von Weizsacker (1980).
5. d'Espagnat (1983).
6. See, for example, Mattuck and Walker (1979), p. 111.
7. Wigner (1967), p. 181.
8. Bohm (1980).
9. Bohr (1963).
10. Schumacher (1984), p. 93.
11. Bohr (1949), p. 222.
12. Leggett (1986).
13. Leggett, *loc. cit.*
14. von Neumann (1955).
15. Ramachandran (1980).
16. Penfield (1976).
17. Schrödinger (1969).
18. Quoted in Rae (1986).
19. Wheeler (1986).
20. Lefebvre (1977).
21. Hofstadter (1980).
22. This is in essence the so-called textbook solution of the measurement problem.
23. This is referred to as the Poincaré-Misra theorem. For a recent review, see Prigogine (1980).
24. Szilard (1929).
25. See Rae (1986); see also Prigogine (1980).
26. I am taking poetic license here. There are a few other attempted solutions to the quantum measurement problem. However, the conclusion stands.

Chapter 7. I Choose, Therefore I Am

1. Baars (1988).
2. Humphrey and Weiskrantz (1967).
3. Humphrey (1972).
4. Shevrin (1980).
5. Sperry (1983).
6. Marcel (1980).

Chapter 8. The Einstein-Podolsky-Rosen Paradox

1. Einstein, Podolsky, and Rosen (1935).
2. Pagels (1982).

3. Bohm (1951).
4. Schrödinger (1948).
5. Aspect, Dalibard, and Roger (1982).
6. Bell (1965).
7. Herbert (1985).
8. For a complete review of all pre-Aspect experiments, see Clauser and Shimony (1978).
9. Bohm claims that there is room for creativity in his theory by virtue of chaos dynamics, see Bohm and Peat (1987); however, as noted in chapter 2, creativity via chaos dynamics is pseudocreativity. Consciousness itself is introduced in Bohm's theory in an arbitrary fashion.
10. Jung (1971), p. 518.
11. *Ibid.*
12. Weinberg (1979).
13. Puthoff and Targ (1976); Jahn (1982).
14. Mermin (1985).
15. Goswami (1986).
16. Grinberg-Zylberbaum et al. (1992).
17. The direct communication requirement makes it impractical to use the subjects' brains as nonlocal telegraphs using Morse code.
18. Monroe (1973).
19. Sabom (1982).
20. Kaufman and Rock (1982).
21. For references to the Russian work, see Jahn (1982).
22. *Ibid.*
23. Mermin (1985).

CHAPTER 9. THE RECONCILIATION OF REALISM AND IDEALISM

1. A similar idea has been proposed by Wolf (1984).
2. Hawking (1990).
3. Wheeler (1986).
4. For a good discussion of the anthropic principle, see Barrow and Tipler (1986).
5. See also d'Espagnat (1983).
6. For a penetrating discussion, see Robinson (1984).
7. Robinson, *loc. cit.*
8. Goswami (1985).
9. In *The Gospel According to Thomas*, Jesus said something similar: "The kingdom [of God] is within you and it is without you." Guillaumont et al. (1959), p. 3.
10. Maslow (1966).

Chapter 10. Exploring the Mind-Body Problem

1. Quoted in Uttal (1981).
2. Such comments are abundant in Skinner's writings. See, for example, Skinner (1976).
3. A good review of identity philosophy can be found in Hook (1960).
4. Berkeley (1965).
5. Sperry (1980).
6. For a very readable introduction to the philosophy of functionalism, see Fodor (1981); Van Gulik (1988).
7. Popper and Eccles (1976).
8. Searle (1980).

Chapter 11. In Search of the Quantum Mind

1. Nikhilananda (1964).
2. Bohm (1951).
3. Harman and Rheingold (1984).
4. *Ibid*, p. 45.
5. *Ibid*, pp. 28–30.
6. *Ibid*, pp. 47–48.
7. Marcel (1980).
8. Selfridge and Neisser (1968).
9. Rumelhart et al. (1986).
10. Posner and Klein (1973).
11. Crick (1978).
12. McCarthy and Goswami (1992).
13. Walker (1970).
14. Eccles (1986).
15. Bass (1975); Wolf (1984).
16. Jahn and Dunn (1986).
17. Feynman (1982).
18. Stuart, Takahashy, and Umezawa (1979).
19. Stapp (1982).
20. Goswami (1990).
21. Jung (1971).
22. In technical language, the idea is that the quantum system of the brain could be the result of Boson condensation. See Lockwood (1989).
23. Orme-Johnson and Haynes (1981).

24. Grinberg-Zylberbaum and Ramos (1987); Grinberg-Zylberbaum (1988).
25. Grinberg-Zylberbaum et al. (1992).
26. See McCarthy and Goswami (in press).
27. Bohr (1963).
28. von Neumann (1955).
29. Hofstadter (1980).

CHAPTER 12. PARADOXES AND TANGLED HIERARCHIES

1. Bateson (1980).
2. Brown (1977).
3. Hofstadter (1980).
4. It is true that the "liar's paradox" stated in this way is not airtight. But it can easily be made airtight by something like, What I am now saying is a lie. However, that is not quite the point. The point is that with our usual language assumptions, "I am a liar" does convey the logical contradiction to most English-speaking adults.
5. Peres and Zurek (1982).

CHAPTER 13. THE "I" OF CONSCIOUSNESS

1. Neumann (1954).
2. Brown (1977).
3. In a recent paper, Mark Mitchell and I have shown that a self-referential generalization of quantum mechanics may be found in a nonlinear Schrödinger equation. The conditioning of a self-referential quantum system follows from the nonlinearity. Mitchell, M. and Goswami, A. In press.
4. Stevens (1964).
5. Attneave (1968).
6. Libet (1979).
7. There may be more intrigue here. In one experiment, Libet and Feinstein used two stimuli: one directly to the skin and the other to an area of the somatosensory cortex that simulates a touch stimulus distinguishable from the skin stimulus. The cortical stimulus was applied first and the skin stimulus a few tenths of a second later. Since both stimuli take about half a second for conscious recognition, it was expected that the cortical stimulus would be reported as the first one sensed. Surprisingly, the subject reported the sensation of the skin stimulus to have occurred first, referring its occurrence to an instant

close to the time of its origin. Libet's explanation is that there is an early time marker in the evoked potential related to the skin stimulus whereas there is no such marker for the cortical stimulus.

Recall (chapter 6) that time's arrow for the manifest world begins with the event of primary collapse. The early time marker of the evoked potential for a skin stimulus may be signaling the primary collapse event, and the patient's backward referral may be due to this fact.

8. Brown (1977).
9. Leonard (1990).
10. Maslow (1968).
11. Eliot (1943).
12. Goswami (1990).
13. Skinner (1962).

Chapter 14. Integrating the Psychologies

1. This chapter is based largely on Goswami and Burns, "The self and the question of free will," unpublished.
2. Husserl (1952).
3. Tart (1975).
4. Rummelhart et al. (1986).
5. Waldrop (1987).
6. Hofstadter (1984), pp. 631–65.
7. Zaborowski (1987).
8. Dollard and Miller (1950).
9. Bandura (1977).
10. Mitchell and Goswami, *op. cit.*
11. Husserl (1952).
12. Maslow (1968).
13. Sartre (1955).
14. Taimni (1961).
15. Dalai Lama (1990).
16. Assagioli (1976).
17. Libet (1985).
18. McCarthy and Goswami (1992).
19. Wilber (1977).
20. Shankara (1975).
21. Sattwa is sometimes wrongly translated as "goodness"; the correct translation is illumination or creativity.
22. Wilber (1979).

CHAPTER 15. WAR AND PEACE

1. Dawkins (1976).
2. Geertz (1973).
3. I am indebted to my colleague, anthropologist Richard Chaney, for many discussions on this subject.
4. Eisler (1987).

CHAPTER 16. OUTER AND INNER CREATIVITY

1. Goswami (1988).
2. Although initially Freud defined libido entirely in terms of the sexual drive, in later writings he seems to use the word to indicate the entire "life force." I use the word *libido* in this more general Freudian sense.
3. Lamb and Easton (1984).
4. Harman and Rheingold (1984).
5. Brown (1977).
6. Bose (1976).
7. Maslow (1968).
8. Krishnamurti (1973).
9. Erikson (1959); Maslow, *loc. cit.*; Rogers (1961).

CHAPTER 17. THE AWAKENING OF BUDDHI

1. Nikhilananda (1964), p. 116.
2. Bateson (1980).
3. Merrell-Wolff (1970).
4. Wallace and Benson (1972).
5. Anand and Chhina (1961).
6. Hirai (1960).
7. Lagmay (1988).
8. Green and Green (1977).
9. Posner (1980).
10. Carrington (1978).
11. Quoted in Joralman (1983).
12. Tagore (1975).
13. A beautiful description of the state of perfect witness can be found in Merrell-Wolff (1973); he called it the state of high indifference.
14. Chaudhury (1981).

15. Nagel (1981).
16. Bly (1977).

Chapter 18. An Idealist Theory of Ethics

1. This chapter is largely based on Goswami, "An idealist theory of ethics," *Creativity Research Journal* (in press).
2. Bloom (1988).
3. Stapp (1985).
4. Kant (1886).
5. Bentham (1976); Mill (1973).
6. Sartre (1980).
7. Orlov (1987); Eddie Oshins, private communication.
8. Garcia (1991).

Chapter 19. Spiritual Joy

1. Aurobindo (1951).
2. Campbell (1968).
3. Ferguson (1980).

BIBLIOGRAPHY

al-Arabi, Ibn. 1976. *Whoso Knoweth Himself*. Translated by W. H. Weir. Gloucestershire, U.K.: Beshara Publications.

———. 1980. *The Bezels of Wisdom*. Translated by R. W. J. Austin. New York: Paulist Press.

Anand, B., and Chhina, G. 1961. "Investigations on yogis claiming to stop their heartbeats." *Indian Journal of Medical Research*. 49:90-94.

Aspect, A.; Dalibard, J.; and Roger, G. 1982. "Experimental test of Bell inequalities using time-varying analyzers." *Physical Review Letters* 49: 1804.

Assagioli, Roberto. 1976. *Psychosynthesis: A Manual of Principles and Techniques*. New York: Penguin.

Attneave, F. 1968. "Consciousness research in men and in man." Paper presented in Burg-Wartenstein Symposium no. 40; preprint University of Oregon, 1968.

Aurobindo, Sri. 1951. *Isha Upanishad*. Pondicherry, India: Aurobindo Ashram Press.

Baars, B. J. 1988. *A Cognitive Theory of Consciousness*. Cambridge: Cambridge University Press.

Bandura, A. 1977. *Social Learning Theory*. Englewood Cliffs, N.J.: Prentice Hall.

Barrow, J. D. and Tipler, F. J. 1986. *The Anthropic Cosmological Principle*. New York: Oxford University Press.

Bass, L. 1975. "A quantum mechanical mind-body interaction." *Foundations of Physics* 5:155–72.

Bateson, G. 1972. *Steps to an Ecology of Mind*. New York: Ballantine.

———. 1980. *Mind and Nature*. New York: Bantam.

Bell, J. S. 1965. "On the Einstein Podolsky Rosen paradox." *Physics* 1:195–200.

Bell, J. S. and Hallett, M. 1982. "Logic, quantum logic and empiricism." *Philosophy of Science* 49: 355.

Bentham, J. 1976. *The Works of Jeremy Bentham*. Edited by J. Bowing. Michigan: Scholarly Press.

Berkeley, G. 1965. *Berkeley's Philosophical Writings*. Edited by D. M. Armstrong. New York: Macmillan.

Berman, Morris. 1984. *The Reenchantment of the World*. New York: Bantam.

Blake, W. 1981. *Poetry and Prose*. Berkeley: University of California Press.

Bloom, A. 1988. *The Closing of the American Mind*. New York: Touchstone.

Bly, R., trans. 1977. *Kabir*. New York: Beacon Press.

Bohm, D. (1951). *Quantum Theory*. Englewood Cliffs, N.J.: Prentice Hall.

_____. 1980. *Wholeness and Implicate Order*. London: Routledge and Kegan Paul.

Bohm, D. and Peat, F. D. 1987. *Science, Order, and Creativity*. New York: Bantam.

Bohr, N. 1949. In *Albert Einstein: Philosopher Scientist*. Edited by P. L. Schilpp. Evanston, Ill.: Library of Living Philosophers.

_____. 1963. *Atomic Physics and Human Knowledge*. New York: Wiley.

Bose, A., trans. 1976. *Later Poems of Rabindranath Tagore*. New York: Minerva.

Brown, Daniel. 1977. In *International Journal of Clinical and Experimental Hypnosis* 25:236–73.

Brown, G. Spencer. 1977. *Laws of Form*. New York: Dutton.

Campbell, Joseph. 1968. *The Hero with a Thousand Faces*. Princeton, N.J.: Princeton University Press.

Capek, M. 1961. *The Philosophical Impact of Contemporary Physics*. Princeton, N.J.: D. Van Nostrand.

Capra, Fritjof. 1975. *The Tao of Physics*. Berkeley: Shambhala.

Carrington, P. 1978. *Freedom in Meditation*. Garden City: Anchor.

Catherine of Genoa. 1979. *Purgation and Purgatory*. Translated by S. Hughes. New York: Paulist Press.

Chaudhuri, H. 1981. *Integral Yoga*. Wheaton, Ill.: The Theosophical Publishing House.

Clauser, J. and Shimony, A. 1978. "Bell's theorem: Experimental tests and implications." *Reports on Progress in Physics* 41:1881.

Colledge and McGinn, trans. 1981. *Meister Eckhart*. New York: Paulist Press.

Crick, F. 1978. "Thinking about the brain." *Scientific American*, Sept., pp. 219–32.

Dalai Lama. 1990. *Ocean of Wisdom: Guidelines for Living*. New York: Harper.

Davies, Paul. 1983. *God and the New Physics*. New York: Simon and Schuster.

Dawkins, R. 1976. *The Selfish Gene*. New York: Oxford University Press.

Descartes, René. 1969. *Philosophical Works*. London: Cambridge University Press.

d'Espagnat, Bernard. 1983. *In Search of Reality*. New York: Springer-Verlag.

DeWitt, B. 1970. "Quantum mechanics and reality." *Physics Today* 23:30.

Dionysius. 1965. *Mystical Theology and the Celestial Hierarchies*. Surrey, U.K.: The Shrine of Wisdom.

Dollard, J. and Miller, N. 1950. *Personality and Psychotherapy*. New York: McGraw Hill.

Dowman, K. 1984. *Sky Dancer: The Secret Life and Songs of Lady Yeshe Tsogyel*. Boston: Routledge and Kegan Paul.

Eccles, John, ed. 1976. *Brain and Conscious Experience*. New York: Springer-Verlag.

———. 1986. "Do mental events cause neural events analogously to the probability fields of quantum mechanics?" *Proceedings of the Royal Society of London* B227:411–28.

Einstein, A.; Podolsky, B.; and Rosen, N. 1935. "Can quantum mechanical description of physical reality be considered complete?" *Physical Review* 47:777–80.

Eisler, R. 1987. *The Chalice and the Blade*. San Francisco: Harper and Row.

Eliot, T. S. 1943. *Four Quartets*. New York: Harcourt Brace Jovanovich.

Erikson, E. 1959. *Identity and the life cycle: Selected papers*. Monograph no. 1, vol. 1. New York: International Universities Press.

———. 1977. *Toys and Reasons*. New York: Norton.

Everett III, H. 1957. *Reviews of Modern Physics* 29:454.

———. 1973. *The Many-Worlds Interpretation of Quantum Mechanics*. Edited by B. DeWitt and N. Graham. Princeton, N.J.: Princeton University Press.

Ferguson, Marilyn. 1980. *The Aquarian Conspiracy*. Los Angeles: J. P. Tarcher.

Feynman, R. P. 1982. "Simulating physics with computers." *International Journal of Theoretical Physics* 21:467–88.

Fodor, J. A. 1981. "The mind-body problem." *Scientific American* 244:114–23.

Garcia, J. D. 1991. *Creative Transformation*. Eugene, Ore.: Noetic Press.

Geertz, C. 1973. *The Interpretation of Cultures*. New York: Basic Books.

Gibbins, P. 1987. *Particles and Paradoxes*. Cambridge: Cambridge University Press.

Gleik, J. 1987. *Chaos*. New York: Viking.

Goddard, D. 1970. *The Buddhist Bible*. Boston: Beacon Press.

Goleman, D. 1986. *Vital Lies, Simple Truths*. New York: Touchstone.

Goswami, A. 1985. "The new physics and its humanistic implications." *Sweet Reason* 4: 3–12.

———. 1986. "The quantum theory of consciousness and psi." *Psi Research* 5:145–65.

———. 1988. "Creativity and the quantum theory." *Journal of Creative Behavior* 22: 9–31.

———. 1989. "The idealistic interpretation of quantum mechanics." *Physics Essays* 2:385–400.

———. 1990. "Consciousness, quantum physics, and the mind-body problem." *Journal of Mind and Behavior* 11:75–96.

Goswami, A. and Burns, J. 1992. "The self and the question of free will." Unpublished.

Green, Elmer and Alice. 1977. *Beyond Biofeedback*. New York: Dell.

Grinberg-Zylberbaum, J. 1988. *Creation of Experience*. Mexico: Instituto Nacional para el Estudio de la Conciencia.

Grinberg-Zylberbaum, J.; Delaflor, M.; Attie, L.; and Goswami, A. 1992. "The EPR paradox in the human brain." To be published.

Grinberg-Zylberbaum, J. and Ramos, J. 1987. "Patterns of interhemispheric correlation during human communication." *International Journal of Neuroscience* 36:41–54.

Guillaumont, A., et al., trans. 1959. *The Gospel According to Thomas*. San Francisco: Harper and Row.

Harman, W. and Rheingold, H. 1984. *Higher Creativity*. Los Angeles: J. P. Tarcher.

Hawking, S. 1990. *A Brief History of Time*. New York: Bantam.

Heisenberg, W. 1930. *The Physical Principles of the Quantum Theory*. New York: Dover.

———. 1958. *Physics and Philosophy*. New York: Harper Torchbooks.

Hellmuth, T.; Zajonc, A. G.; and Walther, H. 1986. "Realizations of the delayed choice experiment." In *New Techniques and Ideas in Quantum Measurement Theory*. Edited by D. M. Greenberger. New York: N. Y. Academy of Science.

Herbert, Nick. 1985. *Quantum Reality*. New York: Doubleday.

Hirai, T. 1960. "Electroencephalographic study of Zen meditation: EEG changes during concentrated relaxation." *Folia Psychiatrica et Neurologica Japanica* 16:76–105.

Hofstadter, Douglas. 1980. *Gödel, Escher, Bach: An Eternal Golden Braid*. New York: Basic Books.

———. 1984. "Waking up from the Boolean dream, or subcognition as computation." In *Metamagical Themas*. New York: Basic Books.

Hook, S., ed. 1960. *Dimensions of Mind*. New York: New York University Press.

Humphrey, N. 1972. "Seeing and nothingness." *New Scientist* 53:682.

Humphrey, N. and Weiskrantz, L. 1967. "Vision in monkeys after removal of the striate cortex." *Nature* 215:595–97.

Husserl, E. 1952. *Ideas: General Introduction to Pure Phenomenology*. Translated by W. R. B. Gibson. New York: Macmillan.

Jahn, Robert. 1982. "The persistent paradox of psychic phenomena: An engineering perspective." *Proceedings of the IEEE* 70:135–70.

Jahn, R. G. and Dunne, B. R. 1986. "On the quantum mechanics of consciousness with applications to anomalous phenomena." *Foundations of Physics* 16:771–72.

James, W. 1958. *Varieties of Religious Experience.* Bergenfield, N.J.: New American Library.

Joralman, D. H. (1983). "When Einstein sat for my mother." *New Age Magazine.* March, p. 40.

Jung, C. G. 1968. *Analytical Psychology: Its Theory and Practice.* New York: Vintage.

———. 1971. *The Portable Jung.* Edited by J. Campbell. N.Y.: Viking.

Jung, C. G. and Pauli, W. 1955. *The Nature and Interpretation of the Psyche.* New York: Pantheon.

Kant, I. 1886. *The Metaphysics of Ethics.* Translated by J. W. Semple. Edinburgh: T. and T. Clark.

Kaufman, L. and Rock, I. 1982. "The moon illusion." *Scientific American,* July, p. 120.

Krishnamurti, J. 1973. *The Awakening of Intelligence.* New York: Avon.

Kuhn, T. S. 1962. *The Structure of Scientific Revolutions.* Chicago: University of Chicago Press.

Lagmay, A. V. 1988. "Science and the Siddhartha: Confluence in two different world views." Paper delivered at the Conference on the Unity of Sciences. Los Angeles, Nov. 24–27.

Lamb, D. and Easton, S. M. 1984. *Multiple Discovery.* Trowbridge, U.K.: Avebury.

Lefebvre, V. 1977. *The Structure of Awareness.* Translated by A. Rappaport. Beverly Hills, Calif.: Sage Publications.

Leggett, A. J. 1986. In *The Lesson of Quantum Theory.* Edited by J. De Boer, E. Dal, and O. Ulfbeck. Amsterdam: North Holland.

Leibniz, G. W. 1898. "Monadology." In *Leibniz: Monadology and Other Philosophical Writings.* Edited and translated by R. Latta. Oxford: Clarendon Press.

Leonard, G. 1990. *The Ultimate Athlete.* New York: North Atlantic.

———. 1978. *The Silent Pulse.* New York: Dutton.

Libet, B.; Wright, E.; Feinstein, B.; and Pearl, D. 1979. "Subjective referral of the timing for a cognitive sensory experience." *Brain* 102:193.

———. 1985. "Unconscious cerebral initiative and the role of conscious will in voluntary action." *The Behavioral and Brain Sciences* 8:529–66.

Lockwood, M. 1989. *Mind, Brain, and the Quantum.* Oxford, U.K.: Blackwell.

London, F. and Bauer, E. 1983. In *Quantum Theory and Measurement.* Edited by J. A. Wheeler and W. Zurek. Princeton: Princeton University Press.

Lowell, J. 1989. "Mr. Eliot's guide to quantum theory." *Physics Today* 42: 46–47.

Marcel, A. J. 1980. "Conscious and preconscious recognition of polysemous words: Locating the selective effect of prior verbal context." In

Attention and Performance VIII. Edited by R. S. Nickerson. Hillsdale, N.J.: Lawrence Erlbaum.

Maslow, Abraham. 1966. *The Psychology of Science*. New York: Harper and Row.

_____. 1968. *Towards a Psychology of Being*. New York: Van Nostrand Reinhold.

_____. 1970. *Motivation and Personality*. New York: Harper and Row.

Mattuck, R. D. and Walker, E. H. 1979. In *The Iceland Papers: Experimental and Theoretical Explorations into the Relation of Consciousness and Physics*. Edited by A. Puharich. Amherst, Wisc.: Essentia Associates.

McCarthy, K. and Goswami, A. In press. "CPU or self-reference? Discerning quantum functionalism and cognitive models of mentation." *Journal of Mind and Behavior*.

Mermin, David. 1985. "Is the moon there when nobody looks? Reality and quantum theory." *Physics Today* 38:38–49.

Merrell-Wolff, F. 1970. *Introceptualism*. Phoenix, Ariz.: Phoenix Press.

_____. 1973. *Pathways Through to Space*. New York: Julian Press.

Mill, J. S. 1973. "On liberty and utilitarianism." In *The Utilitarians*. New York: Anchor.

Milne, A. A. 1926. *Winnie-the-Pooh*. New York: Dutton.

Mitchell, M. and Goswami, A. In press. "Quantum mechanics for observer systems." *Physics Essays*.

Monroe, Robert. 1973. *Journeys Out of the Body*. New York: Doubleday.

Nagel, T. 1981. "What is it like to be a bat?" In *The Mind's Eye*. Edited by D. R. Hofstadter and D. C. Dennett. New York: Basic Books.

Neumann, Eric. 1954. *The Origins and History of Consciousness*. New York: Princeton University Press.

Nikhilananda, Swami, trans. 1964. *The Upanishads*. New York: Harper and Row.

Oppenheimer, J. Robert. 1954. *Science and Common Understanding*. New York: Simon and Schuster.

Orlov, Y. 1987. "A quantum model of doubt." *Proceedings of the N.Y. Academy of Sciences*.

Orme-Johnson, D. W. and Haynes, C. T. 1981. "EEG phase-coherence, pure consciousness, creativity and TM-sidhi experience." *Neuroscience* 13:211–17.

Pagels, Heinz. 1982. *The Cosmic Code*. New York: Simon and Schuster.

Pearle, P. 1984. "Dynamics of the reduction of the state vector." In *The Wave-Particle Dualism*. Edited by S. Diner et al. Riedel: Dordrecht.

Penfield, Wilder. 1976. *The Mystery of the Mind*. Princeton: Princeton University Press.

Penrose, R. 1989. *The Emperor's New Mind*. Oxford, U.K.: Oxford University Press.

Peres, A. and Zurek, W. H. 1982. *American Journal of Physics* 50:807.

Plato. 1980. *Collected Dialogs*. Edited by E. Hamilton and H. Cairns. Princeton, N.J.: Princeton University Press.
Pollard, W. G. 1984. *American Journal of Physics* 52:877.
Popper, Karl, and Eccles, John. 1976. *The Self and Its Brain*. London: Springer-Verlag.
Posner, M. 1980. "Mental chronometry and the problem of consciousness." In *The Nature of Thought: Essays in Honor of D. O. Hebb*. Edited by P. Jusczyk and R. Klein.
Posner, M. I. and Klein, R. 1973. "On the functions of consciousness." In *Attention and Performance*, vol. IV. Edited by S. Kornbloom. New York: Academic Press.
Prigogine, Ilya. 1980. *From Being to Becoming*. San Francisco: Freeman.
Puthoff, H. E. and Targ, R. 1976. "A perceptual channel for information transfer over kilometer distances: Historical perspective and recent research." *Proceedings of the IEEE* 64:329–54.
Rae, A. 1986. *Quantum Physics: Illusion or Reality?* Cambridge: Cambridge University Press.
Ramachandran, S. 1980. In *Consciousness and the Physical World*. Edited by S. Ramachandran and B. Josephson. Oxford: Pergamon.
Ramanan, V. 1978. *Nagarjuna's Philosophy*. New York: Samuel Weiser.
Restak, Richard M. 1979. *The Brain: The Last Frontier*. New York: Doubleday.
Robinson, H. J. 1984. "A theorist's philosophy of science." *Physics Today* 37: 24–32.
Rogers, C. 1961. *On Becoming a Person*. Boston: Houghton Mifflin.
Rummelhart, D. E.; McClelland, J. L.; and the PDP Research Group. 1986. *Parallel Distributed Processing: Explorations in the Microstructure of Cognition*. Vols. 1 and 2. Boston: The MIT Press.
Ryle, Gilbert. 1949. *The Concept of Mind*. London: Hutchinson University Library.
Sabom, M. B. 1982. *Recollections of Death*. New York: Harper and Row.
Sartre, J. P. 1956. *Being and Nothingness*. New York: Philosophical Library.
———. 1980. "Existentialism is a humanism." In *Ethics and the Search for Values*. Edited by L. E. Navia and E. Kelly. New York: Prometheus.
Scholem, G. G. 1954. *Major Trends in Jewish Mysticism*. New York: Schoken Books.
Schrödinger, E. 1948. "The present situation in quantum mechanics." Translated by J. D. Trimmer. *Proceedings of the American Philosophical Society* 124:323–38.
———. 1969. *What Is Life? and Mind and Matter*. London: Cambridge University Press.
Schumacher, J. A. 1984. In *Fundamental Questions in Quantum Mechanics*. Edited by L. M. Roth and A. Inomata. New York: Gordon and Breach.
Searle, J. 1980. "Minds, brains, and programs." *Behavioral and Brain Science* 3:417–24.

Selfridge, O. and Neisser, U. 1968. "Pattern recognition by machine." *Scientific American* 203:69–80.

Shankara. 1975. *Crest Jewel of Discrimination*. Hollywood, Calif.: Vedanta Press.

Shevrin, H. 1980. "Glimpses of the unconscious." *Psychology Today*, April, p. 128.

Skinner, B. F. 1962. *Walden Two*. New York: Macmillan.

———. 1976. *About Behaviorism*. New York: Vintage.

Sperry, R. W. 1980. "Mind-brain interaction: Mentalism, yes; dualism, no." *Neuroscience* 5:195–206.

———. 1983. *Science and Moral Priority*. New York: Columbia University Press.

Squires, E. J. 1986. *The Mystery of the Quantum World*. Bristol, U.K.: Adam Hilger Ltd.

Stapp, H. P. 1977. "Are superluminal connections necessary?" *Nuovo Cimento* 40B:191–99.

———. 1982. "Mind, matter, and quantum mechanics." *Foundations of Physics* 12:363–98.

———. 1985. "Ethics and values in the quantum universe." *Foundations of Physics* 15:35–48.

Stevens, Wallace. 1964. From "The man with the blue guitar." *The Collected Poems*. New York: Knopf.

Stuart, C. I. J. M.; Takahashy, Y.; and Umezawa, M. 1978. "Mixed system brain dynamics." *Foundations of Physics* 9:301–29.

Szilard, L. 1929. "On the decrease of entropy in a thermodynamic system by the intervention of intelligent beings." *Zietschrift Fur Physik* 53:840. In *Quantum Theory and Measurement*. Translated and edited by J. Wheeler and W. Zurek. Princeton, N.J.: Princeton University Press.

Tagore, R. N. 1975. *Fireflies*. New York: Collier.

Taimni, I. K. 1961. *The Science of Yoga*. Wheaton, Ill.: Theosophical Publishing House.

Targ, Russell and Puthoff, Harold. 1977. *Mind-Reach*. New York: Dell.

Tart, Charles. 1975. *The States of Consciousness*. New York: Dutton.

Turing, A. 1964. "Computer machinery and intelligence." In *Minds and Machines*. Edited by A. Anderson. Englewood Cliffs, N.J.: Prentice-Hall.

Uttal, William. 1981. *Psychobiology of the Mind*. New York: Wiley.

Van Gulik, R. 1988. "A functionalist plea for self-consciousness." *The Philosophical Review* 97:149–81.

von Neumann, John. 1955. *The Mathematical Foundations of Quantum Mechanics*. Princeton: Princeton University Press.

von Weizsacker, F. 1980. *The Unity of Nature*. New York: Farrar, Straus, Giroux.

Waldrop, M. 1987. *Man-Made Minds*. New York: Walker.

Walker, E. H. 1970. "The nature of consciousness." *Mathematical Biosciences* 7:131–78.

Wallace, R. and Benson, H. 1972. "The physiology of meditation." *Scientific American*. Feb., pp. 84–90.

Weinberg, S. 1979. *The First Three Minutes*. New York: Bantam.

Wheeler, J. A. 1982. "The computer and the universe." *International Journal of Theoretical Physics* 21:557–72.

_____. 1983. In *Quantum Theory and Measurement*. Edited by J. Wheeler and W. Zurek. Princeton, N.J.: Princeton University Press.

_____. 1986. In *Quantum Measurement Theory*. Edited by D. M. Greenberger. New York: N.Y. Academy of Science.

Wigner, E. P. 1962. In *The Scientist Speculates*. Edited by I. J. Good. Kingswood, Surrey, U. K.: The Windmill Press.

_____. 1967. *Symmetries and Reflections*. Bloomington: Indiana University Press.

Wilber, K. 1977. *The Spectrum of Consciousness*. Wheaton, Ill.: Theosophical Publishing House.

_____. 1979. *No Boundary*. Los Angeles: Center Publications.

Wittgenstein, Ludwig. 1971. *Tractatus Logico-Philosophicus*. London: Routledge and Kegan Paul.

Wolf, Fred Alan. 1981. *Taking the Quantum Leap*. San Francisco: Harper and Row.

_____. 1984. *Starwave*. New York: Macmillan.

Zaborowski, Z. 1987. "A theory of internal and external self-consciousness." *Polish Psychological Bulletin* 18:51–61.

INDEX

CREDITS

Peter W. Atkins, illustration concept of a Fourier wave, from *Quanta: A Handbook of Concepts*, published by Clarendon Press, 1974; Oxford University Press, 1991. Reprinted by permission of the author. T. S. Eliot, excerpt from "Burnt Norton" in *Four Quartets*. Copyright 1943 by T. S. Eliot and renewed 1971 by Esme Valerie Eliot; reprinted by permission of Harcourt Brace Jovanovich, Inc. M. C. Escher, *Drawing Hands*. Copyright © 1948 M. C. Escher/Cordon Art-Baarn-Holland. M. C. Escher, *Print Gallery*. Copyright © 1956 M. C. Escher/Cordon Art-Baarn-Holland. Amit Goswami, from *Quantum Mechanics*. Copyright © 1992 Wm. C. Brown Communications, Inc., Dubuque, Iowa. All Rights Reserved. Reprinted by special permission. Kabir, from *The Kabir Book, A Seventies Press Book*, translated by Robert Bly. Published 1971 by Beacon Press. Reprinted by permission of the publisher. John Lowell, from "Mr. Eliot's Guide to Quantum Theory." *Physics Today*, April 1989, vol. 42, no. 4, p. 47. A. A. Milne, from *Winnie-The-Pooh*. Copyright 1926 by E. P. Dutton, renewed 1954 by A. A. Milne. Used by permission of Dutton Children's Books, a division of Penguin Books USA Inc. Erich Neumann, from *The Origins and History of Consciousness*, transl. R. F. C. Hull. Bollingen Series XLII. Copyright 1954, © 1982 renewed by Princeton University Press. Plates 2, 4, 6 and 9. Reprinted by permission of Princeton University Press. John A. Schumacher, illustration concepts for "The Bohr-Heisenberg Microscope" and "The Eye as a Bohr-Heisenberg Microscope." These concepts are developed further in *Human Posture: The Nature of Inquiry*, by John A. Schumacher, State University of New York Press, Albany, New York, 1989. Wallace Stevens, from *Collected Poems*. Copyright © 1936 by Wallace Stevens and renewed 1964 by Holly Stevens. Reprinted by permission of Alfred A. Knopf, Inc. Rabindranath Tagore, from *Fireflies*, reprinted with permission of Macmillan Publishing Company. Copyright 1928 by Macmillan Publishing Company, renewed 1955 by Rabindranath Tagore.